NEW-TECHNOLOGY FLOWMETERS

New-Technology Flowmeters describes the origin, principle of operation, development, advantages and disadvantages, applications, and frontiers of research for new-technology flowmeters, which include Coriolis, magnetic, ultrasonic, vortex, and thermal. Focusing on the newer, quickly growing flowmeter markets, the book places them in the context of more traditional meters such as differential pressure, turbine, and positive displacement. Taking an objective look at the origins of each flowmeter type, the book discusses the early patents, for each type, and which companies deserve credit for initially commercializing each flowmeter type.

Features:

- Explains the principles of operation of various new-technology flowmeters.
- Describes how each new-technology flowmeter type originated, and who was responsible.
- Discusses the advantages and disadvantages of each new-technology flowmeter type.
- Covers communication protocols, materials of construction, self-diagnostics, and accuracy levels.
- Explores the development of each new-technology flowmeter type over the years, including the roles of companies such as Siemens, ABB, Emerson, Foxboro, KROHNE, and Endress+Hauser.

This book is designed for personnel involved with flowmeters and instrumentation, including product and marketing managers, strategic planners, application engineers, and distributors.

NEW-TECHNOLOGY FLOWMETERS

Volume I

Jesse Yoder

CRC Press is an imprint of the
Taylor & Francis Group, an informa business

First edition published 2023
by CRC Press
6000 Broken Sound Parkway NW, Suite 300, Boca Raton, FL 33487-2742

and by CRC Press
4 Park Square, Milton Park, Abingdon, Oxon, OX14 4RN

CRC Press is an imprint of Taylor & Francis Group, LLC

ISBN: 978-0-367-65542-6 (hbk)
ISBN: 978-1-032-30658-2 (pbk)
ISBN: 978-1-003-13001-7 (ebk)

DOI: 10.1201/9781003130017

Typeset in Times
by MPS Limited, Dehradun

Dedication

This book is dedicated with love to Vicki Tuck. I would also like to dedicate it to the loving memory of my mother and her never-ending prayers for me, and to the loving memory of my father and his never-ending support for me throughout my life.

–Jesse Yoder

Contents

List of Figures

List of Tables

Preface

I have a story to tell. It is not a dry, academic story about equations, definitions, and measurement, even though all those fundamental building blocks play an important role in the story. Neither is it the story of my life, even though my life and experiences are a crucial part of the story. Instead, it is a story that is intended to embody what I have learned about flow and philosophy in 53 years of studying both subjects. It's also a story about what flow is, the many different ways it is measured, and about the many great people and companies that have made it their business to create and market flow measurement devices.

Senator Eugene J. McCarthy

Historic achievements and important inventions are not created in a vacuum. Instead, they are the result of individual decisions made by sometimes ordinary and sometimes extraordinary people. Many people know the story of Paul Revere's midnight ride before the Revolutionary War warning "The British are coming." President John F. Kennedy defused the dangerous Cuban Missile Crisis in 1962 by taking a stand against Nikita Khrushchev's putting missiles into Cuba. John Glenn was the first person to orbit the earth. Anne Frank kept a diary that chronicled her life in hiding from the Nazis from 1942 to 1944. The diary, which became famous after her death, was published in 1947. Senator Eugene McCarthy courageously ran for president in 1968, opposing the Vietnam War. In all these examples, someone made an important decision that influenced the course of history.

In the flow world, companies do not build themselves and inventions do not make themselves. Instead, they often come to be because someone has an idea and tries to make it a reality. Part of this book is devoted to identifying those people who built and created many of the companies and inventions that we take for granted today.

I have spent most of my life studying philosophy and flow. Since I started college, philosophy has been an important part of my life. While some people may find philosophy boring and overly academic, to me it is as exciting as a James Bond movie. I love thinking about how our universe came into being, what it would mean for time to end, what the mind is, and whether points have area. These are some of the topics that I have thought and written about over the years.

HOW IT ALL STARTED

I began studying philosophy in 1969 at the University of Maryland. My first course was a political philosophy course with Dr. Peter Goldstone. It was during this course that I fell in love with philosophy and knew that I wanted to devote my life to philosophy. My honors thesis was an analysis of rules: what they are and under what conditions they exist. I took 63 out of 120 hours of philosophy courses and majored in philosophy of mind. I then went to graduate school at The Rockefeller University, where I took tutorials under Donald Davidson, Joel Feinberg, Harry Frankfurt, and Saul Kripke. After the philosophy program there closed, I transferred to the University of Massachusetts Amherst. I earned my PhD in philosophy in 1984 and wrote about the mind/body problem for my dissertation.

TRANSITION TO PROCESS CONTROL AND MARKET RESEARCH

During the 1980s, I worked as a technical writer for Commercial Union Insurance Cos. and Wang Laboratories. Beginning in 1987, I stated writing software manuals and training guides for the Siemens programmable logic controller (PLC) division. I did this until 1991, when I started doing market research. Market research paid a lot less than technical writing, and so I had to work much longer hours to make ends meet. I worked for five different market research companies until 1998 when I founded Flow Research.

In the early 1990s, I wrote reports on a number of topics, including PLCs, test and measurement, non-destructive test equipment, and numerous other subjects. In 1993, I decided that to do effective market research, I had to understand what I was writing about. This was when I decided to make flow and instrumentation my area of expertise. I began work on my first worldwide flowmeter study for Find SVP in 1993, a study that was published in 1994.

People sometimes ask me how I made the transition from philosophy to flow. My thinking was as follows. My main philosophical interest was philosophy of mind: the nature of the mind, the self, and how mind and body interact. Solving this problem has been a life-long goal. I see the mind as a biological sensor and perceiver. I thought that if I studied electronic and mechanical sensors and transmitters, that perhaps this would help me understand the mind. My study of flow and instrumentation led me to study flow itself, along with flowmeters, which are the devices that measure flow. I discovered that there are many types of flowmeters, and I proceeded to study all of them over the next 29 years.

In addition to flow and flowmeters, I studied and wrote about pressure transmitters and sensors, temperature transmitters and sensors, valves, analytical instruments and level devices. My company Flow Research has published studies on all these topics and more. In many cases having an accurate flow reading requires an accurate temperature and pressure reading. So, the subjects are closely related.

ABOUT THIS BOOK

The main purpose of this book is to provide an in-depth look at the main types of flowmeters. This includes writing about their origin and historical development, along with their theory of operation. While these may seem like straightforward topics, they are not. There is sometimes disagreement about who first invented a technology, or what company first introduced a certain type of flowmeter. While I do not claim to have resolved all these issues in this book, my approach is to try to give my best judgment rather than just to present various alternatives.

My goal in describing the theory of operation of each flowmeter type is to present a clear explanation that is not overly technical. No doubt there are other writers who can present these topics in a more technical way with more equations. My goal is to explain in plain English how the different flowmeters work so the explanation is understandable to those who are not experts in the field. I use equations, photographs, and illustrations as necessary to achieve this goal. In some cases, such as with Coriolis theory, I describe my own perspective while presenting the more traditional view so that the reader can make up his or her own mind.

This book is about new-technology flowmeters. I first created the distinction between new-technology flowmeters and traditional technology (now called conventional) flowmeters in 2000 when doing a market study on the entire flowmeter market. I needed a way to distinguish the more recently introduced and faster growing flowmeters and markets from the more traditional ones that have been around for close to one hundred years or more. These latter types tend to experience slower market growth and are less the subject of ongoing research and development. I introduced this distinction in three articles in *Control* magazine beginning in February 2001. These articles are still posted at www.flowarticles.com.

Initially thermal flowmeters were classified as traditional technology flowmeters, even though they were first introduced in 1974. The main reason was that they did not meet the same accuracy levels as Coriolis, magnetic, ultrasonic, and vortex meters. I was challenged on this subject by John Olin, founder and then president of Sierra Instruments. After six months of discussion with Dr. Olin and other members of the thermal flowmeter community, I was convinced to reclassify them as new-technology flowmeters. I published an article in the October 2003 issue of *Control* magazine explaining my decision. As a result, thermal flowmeters are included in this volume, along with Coriolis, magnetic, ultrasonic, and vortex flowmeters.

There clearly are differences between new-technology and conventional meters. However, some people in the industry find different ways to categorize these differences. I find that some Europeans call new-technology meters "static" meters because they do not have moving parts. Others think of conventional meters as being mechanical meters because most of them do have moving parts. They might refer to new-technology meters as "electronic" meters, although most conventional meters do have electronics attached to them, especially differential pressure, positive displacement, and turbine meters. However, the terms "new-technology" and "traditional technology" or "conventional" have come to be accepted by the industry. I like to think that this is because they mark a genuine distinction between two categories of flowmeter. I have laid out these five criteria in numerous articles over the years.

When discussing the titles of these two volumes with the editors, I found that they prefer the term "conventional flowmeter" to "traditional technology flowmeter." In some ways I find this preferable because I always found "traditional technology flowmeter" to be somewhat wordy, and often substituted the term "traditional flowmeter" or "traditional meter" for it. For this reason, I was happy to go along with the proposed change in terminology. This is why the second volume in this set is called *Conventional Flowmeters*. I have made a corresponding change in terminology at my company Flow Research, and our new studies now refer to conventional flowmeters instead of traditional technology flowmeters.

I selected flowmeters and instrumentation as my main research focus because they have sensors. I saw electronic and mechanical sensors as analogous to the mind, which in many ways is a biological sensor. I see an analogy between electronic sensors and transmitters and our eyes and brains and/or minds. I draw some of these connections in the book *The Tao of Measurement*, which I co-authored with Dick Morley. This book was published by ISA in 2015. I am still working out these connections. I discuss this analogy in more detail in the last chapter of Volume II in this set, *Conventional Flowmeters*.

How does this book differ from my last book *The Tao of Measurement*? There are some similarities, but this book was more broadly focused on measurement, including pressure, temperature, flow, length, and area. My coauthor on *The Tao of Measurement*, Dick Morley, provided commentary on each chapter. This book, by contrast, focuses almost entirely on flow and flowmeters, and also discusses the geometry of flow.

Taking a Closer Look

To better understand the approach taken in this book, it is worthwhile to mention the philosophy of viewpoint pluralism that I developed over the course of a number of years. Simply put, according to viewpoint pluralism, any subject or object is best understood when it is seen from many different points of view. This is an underlying principle that has guided my approach to market research for many years, as well as my attempt to understand many other subjects I have studied. This principle is one of the main reasons why I have discussed flowmeters from so many different perspectives in this book. These perspectives include their origin, the history of their development, their operating principles, their applications, and their frontiers of research. My hope is that this approach will lead to a broader understanding of the flowmeters discussed here rather than just focusing on the theory underlying their operation or even on their applications, important as these perspectives are.

My knowledge of flowmeters has come over 29 years from taking courses, reading books and articles, talking to thousands of professionals in the field, attending and speaking at conferences, and traveling to visit companies in Europe, the Middle East, and Australia. One thing I love about my job is that I learn something new every day, and I never get bored. I have a dedicated staff at Flow Research who helps me do the research and turn the raw data into a presentable study. I am one of those lucky people who loves what he does and sees each day as a new opportunity. I have tried to distill as much of this knowledge as possible into this book. I hope you enjoy it!

Jesse Yoder, PhD
Flow Research, Inc.

Acknowledgments

Even though the ideas in this book mainly address the flowmeter market, my approach to this subject has been guided by a philosophy that I began developing as an undergraduate at the University of Maryland. From there I went on to The Rockefeller University where I wrote three papers a week for the weekly tutorials. Donald Davidson and Joel Feinberg were especially influential on me. Here is where I was able to further develop my writing abilities. This came from writing many analytical philosophy papers and having to defend them in front of professors and fellow students. This is also where I developed my ideas of viewpoint pluralism and circular geometry. At the University of Massachusetts Amherst, Gareth Matthews guided me through the dissertation writing process, and was also a friend to me.

I spent ten years as a technical writer and eight years doing market research before founding Flow Research in 1998. This book incorporates many of the discoveries I made at Flow Research in the past 24 years. Underlying the Flow Research methodology is the philosophy of viewpoint pluralism, and a desire to understand how the input from sensors can be converted into meaningful units of measurement. I have also become very interested in the origin, history, and development of different flowmeter types, and those topics represent the underlying structure of this book.

Belinda Burum has been a wonderful friend to me for 32 years and has served as a partner in building Flow Research. She has been a source of inspiration and insight during this time. Nicole Riordan and Leslie Buchanan have both been a joy to work with. I have worked for many years with the editors of *Process Instrumentation* (formerly *Flow Control*), *Fluid Handling*, and other publications who have helped me reach a broader audience with the results of the data collected by Flow Research. I previously wrote a book called *The Tao of Measurement* with my friend Dick Morley and was very saddened at his recent passing.

Over the years, I have received the support of many very knowledgeable people in the flowmeter industry. Our research could not go on without their support. While this includes hundreds of people, I will mention some who have been especially important and helpful to me. These include Mark Heindselman of Emerson Automation Solutions, Lonna Dickinson of Emerson Rosemount, Mike Touzin of Endress+Hauser, Dick Laan and Andre Boer of KROHNE, Christine Metcalf of Panametrics, a Baker Hughes business, Marlies Slütter of Bronkhorst, Steve Kannengeiszer of Brooks Instrument, and Randy Brown of Fluid Components Int'l. They have been both colleagues and friends to me over the years.

More than anyone else I am grateful to Vicki Tuck for our many years of love, joy, and humor together. Philosophy and flow remain my greatest loves, apart from Vicki.

–Jesse Yoder

1 A Preview of Coming Attractions

OVERVIEW

This is a book about flow and flow measurement. Flow is all around us. Whether it is air flow, water flow, traffic flow, or the flow of gasoline into our cars, it is difficult to escape the impact of flow in our daily lives. Most of us take flow for granted, just as we take gravity and the presence of the sun and moon for granted. Yet it would be difficult to live without flow. If you tried to imagine your life without all those things around us that flow, your life would be either bleak or nonexistent. It is impossible for humans to live without being able to breathe air, and life without water would be short-lived. So even if you have never been inclined to study flow, or even to wonder what it is, your life and the lives of billions of other people on this planet depend on it (Figure 1.1).

Measuring flow is another matter. While we could probably survive without having to measure flow, there are many occasions when flow measurement is important. If you have ever baked a cake, you know the importance of accurate measurement of the liquids involved. You may use a measuring cup, or a glass or cup that is close to eight ounces in size. When you fill up your tank with gas, you want to be assured that you are getting the exact amount of gasoline you are paying for. When buying pre-packaged liquids like milk, you probably assume that the level of the milk in the container accurately reflects the quart or gallon you are buying. Most manufacturers do a good job with this. However, when I compared five of those familiar 500 millimeter water bottles (just over a pint), all marked 16.9 ounces, I was surprised to find that two them had slightly lower water levels than the other three. This means that two had slightly less than 16.9 ounces, unless the other three had slightly more. These differences are small enough that they generally go unnoticed, but it could reflect a slight inaccuracy in the bottling equipment. Almost every measurement is made with a certain tolerance for error.

The idea of how full a container is is actually more noticeable for solids than for liquids. Some cereal boxes and other containers with solid materials contain a disclaimer that reads "Sold by weight, not by volume." This is partly due to the fact that the contents may settle during shipping, and thus the container may not appear to be full. It may also be that the amount of material when weighed simply does not fill up the container, even before shipping. Solids like grains and powders can flow, even though the volume that goes into a container may be determined by a scale rather than by a flowmeter. This book does not discuss the flow of solids, which are not fluids, but nonetheless this is still an important part of the total flowmeter market (Figure 1.2).

DOI: 10.1201/9781003130017-1

FIGURE 1.1 A taste of the future.

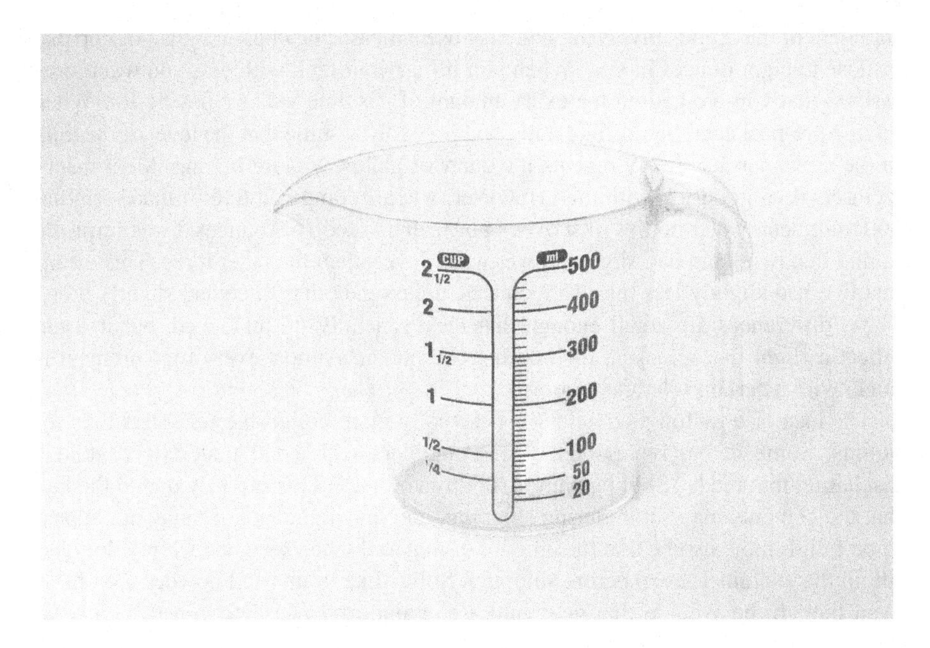

FIGURE 1.2 A measuring cup that includes a milliliter scale.

Marketers play some interesting tricks when it comes to packaging products sold by weight in a container. Coffee is a good example. Coffee used to be sold in one-pound (16 ounce) cans. Then all of a sudden and without notice, the one-pound cans became 13 ounces. The cans looked the same as the one-pound cans, only they contained three ounces less coffee. Apparently this switch happened many years ago. There is an amusing story in the *Deseret News* with a date of May 1, 1989, by Mark Patinkin called "Investigating the Conspiracy: What Happened to the 1-Pound Can?" Mr. Patinkin's story recounts his calls to multiple coffee companies asking why the one-pound can has been reduced to 13 ounces or even 11.5 ounces. The most prevalent answer he got was "Everyone else is doing it." In many cases he was unable to reach anyone to give an answer (Figure 1.3).

Patinkin was able to reach several coffee companies, however. He was able to reach someone at Martinson, which at that time was still committed to the one-pound can. When asked about the future, an associate product manager said, "That decision is I suppose subject to business trends," I was told. "If all the dinosaurs become extinct and you're a dinosaur ..." The idea is that if everyone else is selling the slimmed down cans and you're still selling the one-pound can, you will lose the marketing battle and possibly the marketing war. Apparently the pressure of the slimmed down can eventually got to Martinson too, who decided it did not want to be among the dinosaurs. A quick check on Martinson coffee revealed the availability of a 10.3-ounce can of coffee in 2022.

Perhaps the most creative answer was given by someone named Joan at Folger's. Quoting again from the article:

> "Basically," explained Joan, "Folgers has developed a fast roast method that yields more from each bean. The grounds are more porous like gravel instead of denser like sand, allowing more water to come in contact with the surfaces."

FIGURE 1.3 An 11.3 ounce "pound" of coffee.

> In other words, a new process puffs up the bean, giving it more surface, so when water hits it it yields more coffee. Thirteen ounces allegedly gives you the same bang that 16 used to.

The logic here is pretty obvious. If a company can sell less product for the same or similar price, and people don't notice the difference since the packaging remains the same, the company's profit margin goes up. Mark Patinkin concluded his article by predicting the 3½ quart gallon of milk.

Those large vitamin bottles with many pills snuggled at the bottom, and the bulk of the bottle filled with cotton, raise a similar question. Vitamin D is a good example, because the pills are very small, yet they come in a container much larger than necessary to hold the pills. Here, though, the manufacturers have a better justification. They point out that often a larger bottle is necessary to print all the required text on the bottle. Also, the bottle size has to be large enough so that the "fine print" isn't too small. Another reason for the larger size is that it is more efficient and cost-effective to produce many bottles of a standard size than to produce bottles in a wide range of sizes. In addition, a bottle just large enough to hold 100 vitamin D capsules would potentially get lost on the shelf alongside all the other larger bottles. So this provides more than a marketing justification for those larger vitamin bottles that hold very small pills, tablets, or capsules.

CHAPTER SUMMARY

I would like to think that flow is such a fascinating topic that this book would be of interest to almost anyone. While this may be true, it is also true that the book is likely to be of most interest to those involved in some way with flowmeters, flow measurement, or who are working in industries in which flow measurement is important. The focus of the book is on the main types of new-technology flowmeters. It includes a discussion of their origin, the history of their development including a focus on the companies involved, their principles of operation, and advances in the theory or production of these meters. The book also discusses reasons why people are buying these meters (growth factors), along with applications for them.

This book is the first of a two-part series called "Advances in Flowmeter Technology." For classification and discussion purposes, it is convenient to divide flowmeters into two groups:

1. New-technology flowmeters
2. Conventional flowmeters

This book, which is Volume I of this series, is called *New-Technology Flowmeters*. Volume II is called *Conventional Flowmeters*. The distinction between these two groups of meters is explained in Chapter 2.

A PREVIEW OF UPCOMING CHAPTERS

Chapter 2 explains a number of concepts that are common to all flowmeters and are not unique to one particular type. These include the fundamental flow equation, definitions of flow velocity and flowrate, and the difference between volumetric flow and mass flow. It begins with a definition of flow, explains the process industries, and then discusses the four types of fluid that flowmeters measure: petroleum liquids, gas, non-petroleum liquids, and steam. Air is considered a gas. It also distinguishes between saturated, wet, and supersaturated steam. Even if all these concepts are familiar to you, you might find some of the discussion interesting, such as the review of the fundamental flow equation, the definition of cross-sectional area, the distinction between flow velocity and flowrate, and the definition of mass flow.

Chapter 3 describes a method of selecting flowmeters called the "Paradigm Case Method." When end-users are selecting a flowmeter to use for an application, they might just select the one that is already in place, in case it needs to be replaced. But some applications are new and in some cases there is reason to replace one type of flowmeter with another type. The Paradigm Case Method provides a method with guidelines for selecting a flowmeter that will maximize the possibility of choosing the right flowmeter for a given application.

The remaining chapters focus on particular types of new-technology flowmeters. These include the following types:

- Coriolis
- Magnetic
- Ultrasonic
- Vortex
- Thermal

Chapter 4 discusses Coriolis flowmeters. While Coriolis flowmeters were introduced by Micro Motion in 1977, patents for similar types of meters go back to the 1950s. This chapter describes some of these early patents and examines their relation to the 1977 Micro Motion meters.

Coriolis meters are named after Gustave Coriolis, who observed in the 1840s the effects of the earth's rotation on low and high pressure systems near the equator. Coriolis was not associated with flowmeters, nor did he comment on them. Instead, due to being mentioned in certain patents, his name became associated with Coriolis meters. Chapter 4 examines the extent to which the "Coriolis force" or the "Coriolis effect" plays a role in describing how Coriolis meters actually work.

Chapter 4 examines the principle of operation of Coriolis meters. It describes their development in terms of the companies that introduced Coriolis meters. The chapter discusses the difference between bent tube and straight tube meters. It reviews their advantages and disadvantages, the growth factors for Coriolis meters, and Coriolis meter applications.

Chapter 5 discusses magnetic flowmeters, also called electromagnetic flowmeters. These meters are widely used to measure water flow and the flow of chemicals but cannot be used to measure the flow of petroleum liquids. Chapter 5 describes the initial introduction of magnetic flowmeters in Europe and traces the evolution of magnetic flowmeters as new companies came into the market over time. It discusses the difference between AC (alternating current) and DC (direct current) meters and reviews the advantages and disadvantages of each.

Other topics in Chapter 5 include the growth factors for magnetic flowmeters, along with their applications. Magnetic flowmeters are widely used in the water and wastewater industry, both on the clean water side and for measuring slurries and the flow of wastewater. They are also used in oil and gas to measure chemical injection of fluids into hydraulic fracturing or "fracking" wells. Magnetic flowmeters are especially popular in Europe, where a great deal of water flow measurement occurs.

Chapter 6 discusses ultrasonic flowmeters. Tokyo Keiki introduced the first commercial ultrasonic flowmeters in Japan in 1963. These were clamp-on meters. In 1972, Controlotron released ultrasonic clamp-on meters into the United States. In the 1970s and 1980s, many of the ultrasonic meters on the market were clamp-on meters. Because many users did not fully understand how to install them correctly, ultrasonic meters got a bad reputation during this time. However, with the introduction of inline ultrasonic meters for custody transfer applications in the 1990s and afterward, end-users gained a new respect for ultrasonic meters. Companies like Panametrics and Emerson Process Management entered the market with new innovations. Ultrasonic meters became known for high accuracy and were widely used for custody transfer applications.

This chapter discusses the two main types of ultrasonic flowmeters: transit time and Doppler. Transit time meters are used for clean liquids and gases, while Doppler meters measured the flow of flowstreams with particles. The principle of operation of both meters is described. Over time, manufacturers of transit time improved the ability of transit time meters to measure fluids with impurities, reducing the need for Doppler meters. This chapter discusses the growth factors and applications for both transit time and Doppler meters.

The subject of Chapter 7 is vortex meters. These meters were first introduced in the United States in 1969. They rely on what is called a "bluff body" that is placed in the flowstream. The bluff body generates alternating vortices that are counted by the meter. Flowrate is based on the number of vortices generated by the bluff body in the flowstream.

Some of the main suppliers of vortex flowmeters include Yokogawa, Emerson Automation Solutions, KROHNE, and Endress+Hauser. While vortex meters are widely used to measure the flow of liquid and gas, they do especially well with steam. This is mainly because they can tolerate the high temperatures and pressures associated with steam flow measurement. The chapter describes the principle of operation of vortex meters and explains why end-users are buying them. The chapter also identifies the inventor of multivariable vortex meters, which have become important in mass flow measurement.

Chapter 8 discusses thermal flowmeters, which are almost entirely used to measure gas flow. This chapter includes some of the results of interviews with the founders of thermal flowmeter companies located in or near Monterey, California.

The purpose of these interviews, which were conducted in 2003, was to understand how these companies came into being and the motivation for establishing them.

The interviews focused especially on Fluid Components Int'l, Sierra Instruments, and Kurz Instruments. These companies were formed in the 1960s and 1970s. After these companies were established, new companies were formed, many of which were in California. Other more established companies in Europe and Japan began manufacturing thermal flowmeters at a later time, and these are also described.

There are two main methods by which thermal flowmeters measure flow: constant current and constant power. This chapter describes both methods and provides illustrations that show how these meters work. It also discusses the applications for thermal meters, which include measuring greenhouse gases and flare gas measurement. The advantages and disadvantages of thermal flowmeters are described.

Chapter 9 discusses advances in new-technology flowmeters. This topic is broader than frontiers of research, since some important technological advances have been made in the past 5–10 years that are not necessarily on the current frontiers of research. The chapter looks at three main areas:

1. Custody transfer
2. Advances in flowmeter technology
3. Redundancy

The section on custody transfer focuses on two critical determinants of whether a flowmeter is capable of performing custody transfer measurement:

1. Standards
2. Performance

Coriolis and ultrasonic flowmeters lead the way among new-technology flowmeters in terms of gaining industry approvals for custody transfer measurement. They also can perform at a consistently high level in terms of accuracy and reliability, thereby meeting the performance requirements for custody transfer applications.

A second section in Chapter 9 discusses advances that have been made in each of the five types of new-technology flowmeters. These include meter design, accuracy improvements, improvements in signal processing, better meter tube materials, and better diagnostics. Some of these improvements apply to most types of new-technology meters, though they may take a different form with each type of flowmeter.

The third section describes the growing importance of redundancy in flow measurement. Because some reporting requirements, such as those from the Environmental Protection Agency (EPA), require continuous reporting, redundancy in measurement is more important than ever. Some companies have introduced flowmeters with two or more sensors, and two or more transmitters. Redundancy also plays a role in check metering, and in situations where two flowmeters are run in series. These may be two of a kind, or they may be of different technologies. Check metering is also a form of redundancy. The chapter concludes with a look at the author's patented dual tube flowmeter design, which offers built-in redundancy.

Chapter 10 is called "The Geometry of Flow." It reexamines the definition of flow. Because there is a great deal of measurement in the world of flow, this chapter analyzes some of the fundamental concepts and assumptions that underlie all flow measurement. In particular, it looks at the concepts of point and line, and discusses whether points have area. Most people assume traditional Euclidean geometry when performing flow calculations. I believe that after more than 2,000 years, it is time to take a new look at the validity of Euclidean geometry. This especially includes the idea that the number line is made up of infinitely many dimensionless points. I argue that points have area, that lines have width, and that there is no need for the concept of infinity in the number line if we simply accept the idea that every measurement is made to some degree of precision. Furthermore, infinity times zero = zero.

Most flow measurement is done in round pipes. In order to make this measurement, it is necessary to compute the area of the pipe. Using current geometry, this requires using the formula $A = \pi * r^2$ for the area of a circular pipe. Unfortunately, π becomes necessary because we accept r^2 as our unit of measure. Just as you can't fit a square peg in a round hole, no finite number of squares will fit into a circle.

I propose a new geometry called *circular geometry* that uses a round inch as a unit of measurement to measure the area of round objects. Using squares as a unit of measurement for circles is like using a screwdriver to pound a nail. You need to use the right tool for the type of measurement being made.

The end of Chapter 10 gives a definition of "sensor." Because this is a two-volume set, the second volume will further analyze the flow of information from sensor to transducer to transmitter. It will then explore the analogies between flowmeters as sensors and our own senses including vision. The eye is a biological sensor that passes information to the brain where it is processed. The end result of this process is that we are able to see using two eyes. Somewhat like the dual tube flowmeter described in Chapter 7, our brains take information from both eyes and form a single image of what the eye is focused on. And just like the dual tube meter, there is built-in redundancy in our senses. If one eye fails us, we can still see with the other eye while we get the other eye "fixed."

There is much to learn from studying flowmeters. Whether it is about a new geometry, providing insights into the nature of senses and the mind, or simply experiencing the joy of flow, flowmeters provide a richly rewarding experience.

2 Fundamental Concepts of Flow

While there are many different types of flowmeters, there are some concepts that keep coming up in any discussion of flow measurement. These are concepts like velocity, inner pipe diameter, volumetric flow, mass flow, multivariable flow, flange, insertion meter, etc. This chapter discusses and explains some of these important concepts.

WHAT IS FLOW

"Flow" is a very familiar term, and it is easy to find good uses for this word. Water flows, streams and rivers flow, oil flows, tears flow, time flows onward, current flows, and even traffic flows. What is common to all these examples of flow, and to the many other contexts in which we talk about flow?

The idea of continuous and uninterrupted movement seems central to the idea of flow. But a baseball hit for a home run has continuous and uninterrupted movement into the seats, but it doesn't flow into the seats; it flies. Likewise, a jogger who runs continuously for three miles doesn't flow; he runs or jogs. Perhaps this is because individual persons or objects can't flow. It takes a group of objects in a pattern. We don't say of a fast moving car that it flows, though we might say it can fly. Traffic flow refers to a group of cars moving together in a pattern or "stream." So we talk about the flow of some objects when they move together in a continuous and uninterrupted way in a uniform direction. This is at least a start at a definition (Figure 2.1).

Since flow seems closely connected to continuity, it is worth looking at the idea of continuity. The number line itself is continuous, and yet many mathematicians view the number line as being made up of discrete points. What is confusing about this analysis is that points are conceived of as having no area. This idea seems to be required by the fact that it is always possible to fit another point between any two points. However, if points have no area, meaning they do not have width but are essentially dimensionless, then 1,000 points or one million points will also not have area. Mathematicians compensate for this by using the idea of infinity, arguing that even if points have no area, surely infinitely many of them will have area. But infinity multiplied by 0 is still 0, and adding infinity to dimensionless points does not yield the number line, which is continuous.

Part of the problem with this reasoning is that points are conceived of as making up the line. But if points lie on the line instead of being a part of the line, then the line can have continuity independent of the points. In fact, one way of conceiving of a line is as the path of a point in motion. Likewise, a plane is formed by placing a line in motion.

Flow also exhibits the idea of movement along a path. The key idea behind flow is "continuous and uninterrupted movement along a path in a direction." Since the

DOI: 10.1201/9781003130017-2

FIGURE 2.1 A flowing stream in Boulder, Colorado.

idea of fluid is so closely related to flow, we can include fluids in the definition. To allow for traffic flow but disallow the flying baseball or jogger, we can add the idea of a group of objects moving in a pattern. This brings us to the following definition of flow: "Flow is the continuous and uninterrupted motion of a fluid or a pattern of objects moving uniformly along a path in a direction."

The above definition captures the most common examples of flow, including river flow, the flow of a stream, flow of liquids within a pipe, and open channel flow. An open channel is the opposite of a closed pipe. In open channel flow, the flow moves by gravity, while in closed pipe flow, it is moved under pressure.

Using this definition, it is easier to understand what a flowmeter does. Flow is always in motion, so flowmeters measure the speed of the flow (flow velocity) and compute the amount of flow by using the area or size of the pipe. Flowmeters do not measure water sitting inside a pipe unless it is flowing. Once the fluid goes in motion, it becomes a candidate for measurement.

Some flowmeters place a device into the flowstream to use this the basis for determining flowrate. For example, turbine meters place a rotor into the flowstream and compute flowrate based on the speed of the rotor. Orifice plate meters put a plate with a hole (orifice) in the flowstream to create a pressure difference in the flow. They calculate flow based on this pressure drop. Vortex meters place a thin metal strip called a bluff body that creates vortices; they compute flowrate by counting the number of vortices generated.

Several types of flowmeters measure flow without placing any obstruction into the flowstream. These include magnetic and ultrasonic meters. Magnetic flowmeters create a voltage in the flowstream and measure the voltage with electrodes. Flowrate is proportional to the amount of voltage. Ultrasonic meters transmit ultrasonic waves across the flowstream and back, and use the difference in speed to compute flowrate. Coriolis meters rely on the momentum of the flow as it passes through a bent tube to determine flow. These meters measure very accurately the extent to which the tube is deflected by the flow and compute flow from this value.

FLOWMETERS USED IN THE PROCESS INDUSTRIES

The two-volume series of books called Advances in Flowmeter Technology discusses all the main types of flowmeters used in the process industries. Many of the flowmeters of the type discussed here are also used in the factory and building automation industries, but these applications are not discussed in detail here. Flowmeters used in factory and building automation are generally lower in cost and less sophisticated than the ones used in the process industries. There is a separate group of manufacturers that cater to these industries, and the applications are different from those in the process industries.

DEFINING THE PROCESS INDUSTRIES

It is difficult to give a clear definition of the process industries. Primarily they are the ones concerned with converting raw and bulk materials into products. For example, the food industry takes meat, grains, vegetables, and other potentially edible raw materials and processes them into a form in which they can be consumed by humans and animals. Historically, paper has been made from a variety of materials, including flax, cotton, wood, wheat straw, sugarcane waste, bamboo, linen rags, and hemp. Today most paper is made from wood and recycled paper products. Both food processing and the paper industry are examples of process industries.

By contrast with the process industries, the factory and building automation industries work with discrete things that can be put together to form finished products. The automotive industry takes physical objects like tires, spark plugs, radiators, steering columns, doors, and many other parts and assembles them into a marvelous and complex automobile that is used to transport us where we want to go. The building industry uses concrete, wood, steel, masonry, and stones, along with many other materials, to create houses, industrial buildings, factories, and other structures that are used as homes, offices, and manufacturing locations.

A different way to understand the process industries is simply by listing them. A list of the main process industries included in this book is as follows:

- Oil and gas
- Refining
- Chemical
- Food and beverage
- Pharmaceutical
- Pulp and paper
- Metals and mining
- Power
- Water and wastewater

There are some additional process industries that are not as large as the above ones such as plastics, ceramics, rubber, tobacco, and textiles that are sometimes included in a category called "Other."

One main feature of the process industries is that they all involve the use of fluids to accomplish their purposes. A fluid is anything that flows. The fluids mainly discussed in this book are:

- Petroleum liquids
- Natural gas and industrial gases
- Non-petroleum liquids
- Steam

It is natural to associate the word "fluid" with "liquid," but gases and steam are also fluids. I once asked the editor of *Fluid Handling* magazine if I could submit an article on gas flow measurement. He said "How can we accept that? Our magazine is called 'Fluid Handling.'" I pointed out to him that both gas and steam are fluids, and that his magazine is not called "Liquid Handling." I got the article published, and since that time *Fluid Handling* has covered gas and steam flow as well as liquid flow and measurement.

Petroleum Liquids

Petroleum liquids are hydrocarbons that were formed from the remains of animals and plants that lived millions of years ago. Hydrocarbons are molecules of hydrogen and carbon in different combinations. In the oil and gas industry, wells are drilled deep into the earth for oil and gas, and both of them are brought to the surface. Once there, they are separated from each other and from any water that is present. Then they go through test separators and production separators, which strip off extraneous fluids. After being "purified" in this manner, the oil is sent via truck, train, pipeline, or ship to a refinery. At the refinery the oil is divided into its component parts through distillation. The result is the refined fuels that we use for transportation, including gasoline, motor oil, diesel fuel, kerosene fuel, heating oil, and many other refined petroleum products. From the refinery, these products are either stored, put in containers, or sent by pipeline, truck, or other means to their point of use (Figure 2.2).

Natural Gas and Industrial Gases

Gas that comes to the surface as part of the drilling process is called natural gas. The largest component of natural gas is methane, but it contains some other gases as well such as ethane, butane, propane, and pentane. One it reaches the surface, it follows a process similar to that of crude oil. It also goes through test separators and production separators, which strip out extraneous fluid. Once the gas leaves the oilfield, it goes to a gas processing plant through a gathering system of pipelines. The processing plant is similar in function to an oil refinery. Gas processing plants remove contaminants from the gas such as carbon dioxide, water, hydrogen sulfide, and mercury. Some of these products that are contaminants in natural gas have

FIGURE 2.2 A wellhead, commonly called a Christmas tree on a well near Traverse City, Michigan.

economic value and are removed and sold separately. The main output of a natural gas plant is high quality natural gas that can be used as a fuel by residential, commercial, and industrial customers, or that can be used as feedstock for chemical synthesis. Natural gas is typically delivered by natural gas pipelines. This can be a long process involving thousands of miles of pipeline in which the gas goes to a distribution company that delivers it to the end-users.

Besides natural gas, industrial gases are manufactured for use in industry. Industrial gases typically are stored and transported in a cylinder. Industrial gases are used in a wide range of industries including oil and gas; chemical, food, and beverage; pharmaceutical, power, water, and wastewater; and other industries. Common industrial gases include hydrogen, carbon dioxide, oxygen, helium, nitrogen, and argon.

Non-petroleum Liquids

The main fluids involved in the oil and gas industry are crude oil, refined fuels, natural gas, and other gases. However, non-petroleum liquids such as water and other chemicals also play an important role at many points in this industry. Water is stripped from petroleum liquids after they leave the wellhead, and it is also removed from natural gas and associated gases. On the positive side, water is an integral part of the refining process. According to a report from the Department of Energy, a typical refinery uses about 1.5 barrels of water to process one barrel of crude oil. Steam is also involved in the refinery distillation process. This means that all the main fluid types: oil, gas, water, and steam are an integral part of the oil and gas industry.

Non-petroleum liquids are also widely used in the other process industries. In the chemical industry, liquid chemicals are used as additives in the process of making many chemicals. Liquids are widely involved in brewing beer and in making soda for drinking. In the pulp and paper industry, the pulp is in liquid form. Power plants rely on heated water to generate steam, which turns turbine blades to create power. And the main purpose of the water and wastewater industry is either to create clean water for human consumption or to treat wastewater so that it can either be reused or safely disposed of.

STEAM

Anyone who has boiled water knows what steam is. Steam is the vapor that rises from the boiling water in cloud-like form. This is considered to be water in its gaseous state. This concept is sufficient for most applications in daily life; however, in industrial environments it is customary to distinguish three types of steam:

1. Saturated steam
2. Wet steam
3. Supersaturated steam

These three types of steam depend on the pressure and temperature of the steam. Saturated steam occurs at the boiling point of water (212°F or 100°C). Once the boiling point is reached and further heat is added, the water vaporizes and converts to steam. At this point, the temperature of the saturated steam is equal to the boiling point of the water that it vaporizes from. Saturated steam is invisible to the human eye; however, as it rises in the air, its temperature drops and it forms water droplets that take the form of wet steam. These are the clouds of water vapor that are commonly known as steam. Saturated steam is also called dry steam. Saturated steam is used as a heat source for reactors, heat exchangers, and other heat transfer equipment.

Wet steam is a mixture of water droplets and steam. When the percent of water droplets exceeds 5 percent, the vapor is a two-phase fluid and is considered to be wet steam. Wet steam is characterized by its quality based on the percent of water droplets in the steam. For example, steam that is 10 percent water and 90 percent steam is called 90 percent quality steam. Wet steam can cause corrosion in turbine blades and heat transfer equipment (Figure 2.3).

Superheated steam is steam that has a higher temperature than saturated steam but that has maintained its pressure. Superheated steam acts as a gas is highly sensitive to temperature and pressure measurements. It is used for energy systems, cleaning and surface technologies, and chemical reaction processing.

FUNDAMENTAL CONCEPTS

Regardless of how flow is measured, there are certain concepts that are important to understand. One of these is the fundamental flow equation:

$$Q = A * v$$

FIGURE 2.3 Steam flows from a boiler in India.

In this equation:

- Q = volume per unit time
- A = cross-sectional area of the pipe at a specific point
- V = average velocity of the flow at that specific point

The letter Q has traditionally been used to symbolize flow. It was introduced by French scientists in the 19th century. The letter Q comes from the French word "Quantité," which means "amount" or quantity.

CROSS-SECTIONAL AREA

In this case, we are considering flow within a closed pipe. When considering the area of a pipe, it is important to consider the inside diameter of the pipe, since that is the area the flow passes through. The difference between the inside diameter and the outside diameter of a pipe is simply the width of the pipe wall multiplied by 2. So one way to determine a pipe's inside diameter if the outside diameter is known is to take the outside diameter, measure the width of the pipe wall, and subtract a value that is twice the width of the pipe wall.

Since pipes are round, computing the cross-sectional area involves using the formula for the area of a circle. The cross-sectional area of a pipe is the area of a circle that is seen when looking at the end of the pipe. This area is determined from the inside diameter of the pipe. The inside diameter of a pipe is the distance of a straight line from a point on the inner wall of one side of the pipe through the center of the pipe to an opposite point on the inner wall on the other side of the pipe.

Once the inner diameter of the pipe is known, the cross-sectional area of the pipe can be calculated. The formula for the area of a circle is:

$$\pi * r^2$$

where r is the radius of the inner diameter. Since the radius of the inner pipe diameter is $D/2$, where D is the inner diameter, the equation for the cross-sectional area of the pipe becomes:

$$A = \pi * (D/2)^2 = \frac{1}{4}\pi * D^2$$

This means, for example, that the cross-sectional area of a pipe with an inner diameter of two inches is equal to π, or 3.14 square inches.

FLOW VELOCITY

Flow velocity is a measurement of how fast the flow is moving per unit of time. A common Imperial (American) unit for measuring flow velocity is feet per second. A common SI (metric) way to measure flow velocity is meters per second.

VOLUMETRIC FLOWRATE

Q, or volumetric flowrate, is a measurement of the total volume of fluid flowing through a pipe per unit of time. Typical units for volumetric flowrate are liters per second, cubic centimeters per minute, cubic meters per second, cubic meters per hour, cubic feet per second, gallons per hour, and other units.

MASS FLOWRATE

The mass of a body measures the amount of material it contains and causes it to have weight in a gravitational field. Mass flow can be calculated from volumetric flow if its temperature and pressure are known. Mass flowrate W can be determined from the volumetric flowrate Q and fluid density ρ:

$$W = Q * \rho$$

Mass flow is especially important in measuring gas flow, since the number of molecules in a gas varies with pressure and temperature. It is less important in measuring liquids because liquids are nearly incompressible. Coriolis and thermal flowmeters measure mass flow directly, while multivariable flowmeters measure mass flow indirectly. It can also be said that multivariable flowmeters infer mass flow by measuring volumetric flow and then using pressure and temperature values to compute mass flow. Examples of multivariable flowmeters are multivariable vortex and multivariable differential pressure (DP) meters.

3 The Paradigm Case Method of Selecting Flowmeters

OVERVIEW

This chapter gives a general overview of all flow measurement technologies and flowmeter selection. It covers new-technology and conventional flowmeters as groups, covers each individual type of flowmeter technology, with a paradigm case for each, and then lays out a paradigm case method for flowmeter selection.

INTRODUCTION TO FLOWMETER TYPES AND FLOWMETER SELECTION

One of the most interesting developments in the flowmeter market today is the battle between the newer flow technologies and the conventional flowmeters. New-technology flowmeters include Coriolis, magnetic, ultrasonic, vortex, and thermal meters. The conventional flow technologies include DP – and primary elements – positive displacement, turbine, open channel, and VA flowmeters. While there is a general trend toward the new-technology meters and away from the conventional flowmeters, the rate of change varies greatly by industry and application.

When users select flowmeters today, they are faced with a variety of choices. Not only are many technologies available, but there are many suppliers for each technology. When ordering replacement meters, users often replace like with like. This is one reason why DP flowmeters still have the largest installed base of any type of flowmeter. In other cases, users need to select meters for new plants, or for new applications within existing plants. Users also sometimes replace one type of flowmeter with another type. How should this selection be made?

This chapter addresses the issue of flowmeter selection by examining the operating principles, advantages, and limitations of the new-technology and conventional types of flowmeters. It also presents a step-by-step method of flowmeter selection that takes these technology factors into account, along with application, performance, cost, and supplier capabilities. This decision-making process is called the "paradigm case" method of flowmeter selection. Because the first step in the paradigm case method of flowmeter selection involves selecting flowmeters whose paradigm case application best matches a particular field of application, the paradigm case application is identified for each type of flowmeter. A "paradigm case" application is one in which the conditions are optimal for the operation of that type

DOI: 10.1201/9781003130017-3

of flowmeter. The paradigm case application for each flowmeter type is determined by the physical principle by which the flowmeter senses flow. The paradigm case is included in the discussion of each flowmeter type.

NEW-TECHNOLOGY FLOWMETERS

New-technology flowmeters acquired this name in part because they represent technologies that have been introduced more recently than DP flowmeters. Most of the new-technology flowmeters came into industrial use in the 1960s and 1970s, while the history of DP flowmeters goes back to the early 1900s. Each new-technology flowmeter is based on a different physical principle and represents a unique approach to flow measurement.

New-technology flowmeters share several characteristics.

1. They were introduced after 1950.
2. They incorporate technological advances that avoid some of the problems inherent in earlier flowmeters.
3. They are more the focus of new product development efforts by the major flowmeter suppliers than conventional flowmeters.
4. Their performance, including criteria such as accuracy and repeatability, is at a higher level than that of most conventional flowmeters.
5. They are relatively quick to incorporate advanced in communication protocols such as HART, Foundation Fieldbus™, Profibus®, Modbus, and other communication protocols for networking purposes.
6. They either do not rely on placing a device into the flowstream to make a flow measurement, or any device placed in the flowstream for measurement purposes causes minimal interference with the flow and minimal pressure drop.

Those flowmeters that incorporate newer technologies are classified here as new-technology flowmeters. This category includes Coriolis, magnetic, ultrasonic, vortex, and thermal flowmeters. These meters were all introduced after 1950. Magnetic flowmeters were first introduced in Holland in 1952. Tokyo Keiki first introduced ultrasonic meters in Japan in 1963. Eastech brought out vortex flowmeters in 1969, while Coriolis meters came onto the market in 1977. The most widely known type of thermal flowmeters were developed in the mid-1970s, although some thermal meters were on the market as early as 1959.

Just as flowmeters that incorporate new technologies are classified as new-technology meters, flowmeters that incorporate more conventional technologies are classified as conventional flowmeters. These include DP, turbine, positive displacement, open channel, and VA meters. As a group, these meters have been around longer than the new-technology meters. They generally have higher maintenance requirements than new-technology flowmeters. And while suppliers do continue to bring out enhanced conventional flowmeters, they are less the focus of new product development than new-technology meters.

The history of DP meters goes back to the early 1900s, while the beginnings of the turbine meter go back to at least the mid-1800s. Many of the problems inherent in DP meters are related to the primary elements they use to measure flow. For example, orifice plates are subject to wear, and can also be knocked out of position by impurities in the flow stream. Turbine and positive displacement meters have moving parts that are subject to wear. The accuracy levels of open channel and VA meters are significantly lower than those of new-technology flowmeters.

Coriolis Flowmeters

Coriolis flowmeters are named after the French mathematician Gustave Coriolis. In 1835, Coriolis showed that an inertial force must be taken into account when describing the motion of bodies in a rotating frame of reference. The earth is commonly used as an example of the Coriolis force. Imagine an object thrown from the North or South Pole aiming in a straight line toward a spot on the Equator. However, the earth's rotation also affects the object and the object will deviate from the intended straight-line path resulting in a curving path.

Coriolis flowmeters are composed of one or more vibrating tubes, usually bent. The fluid to be measured passes through the vibrating tubes. The fluid accelerates as it passes toward the point of maximum vibration and decelerates as it leaves this point. The result is a twisting motion in the tubes. The degree of twisting motion is directly proportional to the fluid's mass flow. Position detectors sense the positions of the tubes. While most Coriolis flowmeter tubes are bent, and a variety of designs are available, some manufacturers have also introduced straight-tube Coriolis flowmeters.

Straight-tube flowmeters work on the same principle as do the bent-tube meters. The inertia of the fluid causes acceleration of fluid particles in the first half of the flowmeter, and deceleration of fluid particles in the second half of the meter. This inertia generates a Coriolis force that slightly distorts the measuring tube. This distortion, which is detected by special sensors, is proportional to mass flowrate. Because the oscillatory properties of the measuring tube vary with temperature, temperature is continuously measured so that the necessary adjustments can be made in the measurement.

It is often said that Coriolis flowmeters measure mass flow "directly," unlike other flowmeters that calculate mass flow by using an inferred density value. Volumetric flow (Q) is calculated by multiplying the cross-sectional area of a pipe by the fluid's average velocity. Mass flow is then determined by multiplying volumetric flow (Q) by the density of the fluid. Some multivariable flowmeters measure the pressure and temperature of the process fluid, and then use these values to infer fluid density. Mass flow can then be calculated.

Coriolis flowmeters are used on both liquids and gases. While they are highly accurate, they are limited in size to six inches or less, with a few exceptions. Coriolis flowmeters have relatively high initial cost, although some models are now available in the $3,000 range. Their high initial cost is offset by their normally low maintenance costs. Coriolis meters can handle some fluids with varying densities that cannot be measured by other flowmeters.

Paradigm Case Application

The paradigm case application for Coriolis meters is with *clean liquids and gases flowing sufficiently fast to operate the meter through pipes two inches or less in diameter*. The primary limitation on Coriolis meters is size since they become quite unwieldy in sizes over two inches. While three-inch and four-inch Coriolis meters are available, conditions are not ideal for meters of these diameters, due to the required size of the meter. Some low-pressure gases do not have sufficient density to operate Coriolis flowmeters. One advantage of Coriolis meters is that the same flowmeter can be used to measure different types of fluids, including fluids of different densities. Coriolis meters can measure the mass flow of slurries and dirty liquids, but these fluids should be measured at relatively low flowrates to minimize meter wear.

Magnetic Flowmeters

Magnetic flowmeters use Faraday's law of electromagnetic induction. According to this principle, a voltage is generated in a conductive medium when it passes through a magnetic field. This voltage is directly proportional to the density of the magnetic field, the length of the conductor, and the velocity of the conductive medium. In Faraday's law, these three values are multiplied together, along with a constant, to yield the magnitude of the voltage.

Magnetic flowmeters use wire coils mounted onto or outside of a pipe. Current is then applied to these coils, generating a magnetic field inside the pipe. As the conductive liquid passes through the pipe, a voltage is generated and detected by electrodes, which are mounted on either side of the pipe. The flowmeter uses this value to compute the flowrate.

Magnetic flowmeters are used to measure the flow of conductive liquids and slurries, including paper industry pulp slurries and black liquor (produced during wood to paper pulp process). Their main limitation is that they cannot measure hydrocarbons (which are nonconductive), and hence are not widely used in the petroleum industry. "Magmeters," as they are often called, are highly accurate and do not create pressure drop. Their initial purchase cost is relatively high, though most magmeters are priced lower than equivalent Coriolis meters.

Paradigm Case Application

The paradigm case application for magmeters involves *conductive fluids flowing through a full pipe that do not contain materials that damage the liner or coat the electrodes*. The most obvious and serious limitation on the use of magnetic flowmeters is that they only work with conductive fluids. This principle works with conductive liquids but not with gases or steam, so magmeters only work with conductive liquids. Because they compute flowrate based on velocity times area, accurate readings require that the pipe be full. In addition, electrode coating and liner damage can degrade the accuracy of magnetic flowmeters.

Ultrasonic Flowmeters

Ultrasonic flowmeters were first introduced for industrial use in 1963. There are two main types of ultrasonic flowmeters: transit time and Doppler.

Transit time ultrasonic meters have both a sender and a receiver. They send an ultrasonic signal across a pipe at an angle and measure the time it takes for the signal to travel from one side of the pipe to the other. When the ultrasonic signal travels with the flow, it travels faster than when it travels against the flow. The ultrasonic flowmeter determines how long it takes for the signal to cross the pipe in one direction, and then determines how long it takes the signal to cross the pipe in the reverse direction. The difference between these times is proportional to flowrate. Transit time ultrasonic flowmeters are mainly used for clean liquids.

Doppler ultrasonic flowmeters also send an ultrasonic signal across the pipe. However, instead being sent to a receiver on the other side of the pipe, the signal is reflected off particles traveling in the flowstream. These particles are traveling at the same speed as the flow. As the signal passes through the stream, its frequency shifts in proportion to the mean velocity of the fluid. A receiver detects the reflected signal and measures its frequency. The meter calculates flow by comparing the generated and detected frequencies. Doppler ultrasonic flowmeters are used with dirty liquids or slurries.

One significant development for ultrasonic flowmeters was the approval by the American Gas Association (AGA) in June 1998 of criteria for using ultrasonic flowmeters for custody transfer of natural gas. This approval gave a major boost to the ultrasonic flowmeter market in the oil and gas production and transportation industry. Only multipath flowmeters are approved for use in custody transfer. Most multipath flowmeters use four to six different paths of ultrasonic signals to determine flowrate.

Multipath ultrasonic flowmeters use multiple pairs of sending and receiving transducers to determine flowrate. The transducers alternate in their function as sender and receiver over the same path length. The flowrate is determined by averaging the values given by the different paths, providing greater accuracy than single path meters.

Paradigm Case Application

Ultrasonic flowmeters can handle liquids and gases. They can be affected by swirl. The paradigm case application for transit time ultrasonic flowmeters is *clean, swirl-free liquids and gases of known velocity profile.* The need for high accuracy may require the use of a multipath meter. The most important constraint on ultrasonic flowmeters is that the fluid be clean, although today's transit time meters can tolerate some impurities. A single-path ultrasonic meter calculates flowrate based on a single path through the pipe, making it susceptible to flow profile aberrations. Multipath flowmeters are more accurate, since they use multiple paths to make the flow calculation, usually from four to six paths. Ultrasonic flowmeters are available as inline and clamp-on models. The paradigm case application for clamp-on models requires taking characteristics of the pipe into account, as well as fluid characteristics.

VORTEX FLOWMETERS

Vortex flowmeters employ a principle called the von Karman effect. According to this principle, flow will alternately generate vortices when passing by a bluff body. A bluff body has a broad, flat front. In a vortex meter, the bluff body is a piece of material with a broad, flat front that is mounted at right angles to the flowstream. Flow velocity is proportional to the frequency of the vortices. Flowrate is calculated by multiplying the area of the pipe times the velocity of the flow.

In some cases, vortex meters require the use of straightening vanes or a specified length of straight upstream and downstream piping to eliminate distorted flow patterns and swirl. Low flowrates present a problem for vortex meters because low flowrates generate vortices irregularly. The accuracy of vortex meters is from medium to high, depending on model and manufacturer. In addition to liquid and gas flow measurement, vortex flowmeters are widely used to measure steam flow.

Paradigm Case Application

Paradigm case applications for vortex meters are *clean, low viscosity, swirl-free fluids flowing at medium to high speed.* Ideal conditions for vortex flowmeters include medium to high flowrates because formation of vortices is irregular at low flowrates. Since swirls can interfere with the accuracy of the reading, the stream should be swirl-free. Any corrosion, erosion, or deposits that affect the shape of the bluff body can shift the flowmeter calibration, so vortex meters work best with clean liquids. Vortex meters also work best with low-viscosity fluids.

THERMAL FLOWMETERS

Thermal flowmeters were developed in the late 1960s and early 1970s as an offshoot of research into air velocity profile and turbulence research. This research, conducted by Dr. Jerry Kurz and Dr. John Olin, used hot-wire anemometers. These anemometers consisted of a heated, thin-film element, but they were too fragile for industrial environments. Kurz and Olin collaborated to form a company called Sierra Instruments in 1973. They developed a thermal flowmeter that was more rugged and hardened than the hot-wire anemometers. When they split up in 1977, Olin stayed with Sierra and Kurz formed Kurz Instruments. Both companies are located in Monterey, California.

Fluid Components International (FCI) took a different approach to developing thermal flowmeters. FCI was founded in 1964 by Mac McQueen and Bob Deane. The company developed flow switches to detect the flow of oil through pipes in the oil patch. Although these thermal switches were not flowmeters, they formed the basis of what later became flowmeters. In 1981, FCI put more sophisticated electronics on its switches, creating thermal flowmeters for gas flow measurement.

Thermal flowmeters work by introducing heat into the flowstream and measuring how much heat dissipates, using one or more temperature sensors. There are two different methods for doing this.

One method is called the constant temperature differential. Thermal flowmeters that use this method have two temperature sensors. One is a heated sensor and the

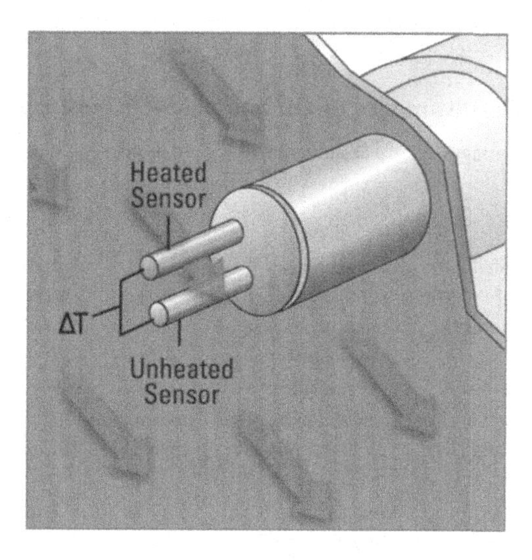

FIGURE 3.1 Thermal mass flowmeter technology. (Source: Graphic courtesy of Fluid Components Int'l.)

other sensor measures the temperature of the gas. Mass flowrate is calculated based on how much electrical power is required to maintain a constant temperature difference between the two temperature sensors (Figure 3.1).

The second method is called a constant current method. Under this method, thermal flowmeters also have two sensors: one heated sensor and one that measures the temperature of the flow stream. Power to the heated sensor is kept constant. Mass flow is measured based on the difference between the temperature of the heated sensor and the temperature of the flowstream.

Both methods make use of the principle that higher velocity flows result in greater cooling. Both compute mass flow by measuring the effects of cooling on the flow stream.

Paradigm Case Application

The paradigm case application for thermal flowmeters is *clean, pure, flowing gases of known heat capacity or clean mixtures of pure gases of known composition and known heat capacity under temperature conditions of 100°C or less*. Thermal flowmeters have limited application to liquids, and they do not work for measuring steam flow.

SUMMARY TABLES OF ADVANTAGES, DISADVANTAGES AND PRINCIPLES OF OPERATION

Table 3.1 summarizes the advantages and disadvantages of DP and new-technology flowmeters, along with some additional application criteria. The characteristics of the main types of DP flowmeters and various primary element types are also included in this table.

TABLE 3.1
Advantages and Disadvantages of DP and New-Technology Flowmeters

Flowmeter Type	Advantages	Disadvantages	Liquid, Steam, or Gas	Pipe Size	Comment
DP – orifice plate	Low initial cost; ease of installation; well understood	Limited range; permanent pressure drop; uses square root method to calculate flowrate	Liquid, steam, gas	½ inch and up	Most conventional meter
DP – Venturi tube	Suitable for clean and dirty liquids	Can be unwieldy and difficult to install due to size	Liquid, gas	2–30 inches	Greater size for gas
DP – pitot tube	Low cost; virtually no pressure drop	Low accuracy; limited sampling	Liquid, gas	> 1 inch	Measures only at a single point
DP – averaging pitot tube	More accurate than single pitot tube; virtually no pressure drop	Limited range; not suitable for dirty fluids	Liquid, gas	> 1 inch	Available as Annubar
DP – flow nozzle	Good for high velocity fluids; handles dirty fluids better than orifice plate	High initial cost; difficult to remove for inspection and cleaning	Steam	2–30 inches	Used for steam applications
Coriolis	High accuracy; measures mass flow directly	Expensive and unwieldy in pipe sizes over six inches; high initial cost; sensitive to vibration	Liquid, gas	1/16–16 inches	Measures mass flow directly
Magnetic	Obstruction-less; high accuracy; no pressure drop	Cannot meter nonconductive fluids (e.g., hydrocarbons); relatively high initial cost; electrodes subject to coating	Liquid	1/10–100 inches	Limited use in the petroleum industry because it does not meter hydrocarbons
Ultrasonic – single path transit time	High accuracy; nonintrusive	High initial cost; requires clean fluids; sensitive to swirl	Liquid, gas	½ inch and up	Used for check metering applications
Ultrasonic – multipath transit time	High accuracy; nonintrusive	High initial cost; requires clean fluids; sensitive to swirl	Liquid, gas	4–36 inches	Approved for custody transfer of natural gas
Ultrasonic – Doppler	Operates on dirty liquids; nonintrusive	Low-medium accuracy	Liquid	½ inch and up	Limited accuracy but one of few meters designed for dirty liquids

TABLE 3.1 (Continued)
Advantages and Disadvantages of DP and New-Technology Flowmeters

Flowmeter Type	Advantages	Disadvantages	Liquid, Steam, or Gas	Pipe Size	Comment
Vortex	Medium-high accuracy	Vibration can affect accuracy; limited industry approvals	Liquid, steam, gas	½–12 inches	Widely used for steam measurement
Thermal	Handles large line sizes	Lacks industry approvals	Gas, limited liquids	¼ inch and up	Almost entirely used for gas flow measurement

Table 3.2 summarizes the operating principles of new-technology flowmeters. Because users often need to choose between new-technology and some type of DP flowmeters, the table also summarizes the main types of DP flowmeters.

CONVENTIONAL FLOWMETERS

Despite the growth of new-technology flowmeters such as magnetic, Coriolis and ultrasonic over the years, conventional flowmeters are holding their own. Many users are still selecting DP, turbine, positive displacement, and other conventional flowmeters as their flowmeter solutions. This section describes the conventional flowmeter market and explains why this market retains a large portion of flowmeter market share even in the face of these competing technologies.

Conventional flowmeters share the following characteristics:

1. They were introduced before 1950.
2. They are relatively slow to incorporate technological advances that avoid some of the problems inherent in earlier flowmeters.
3. They are less the focus of new product development by the major flowmeter suppliers than new-technology flowmeters.
4. Their performance, including accuracy and repeatability, is generally not at as high a level as that of new-technology flowmeters.
5. They are relatively slow to incorporate advances in communication protocols, such as HART, Foundation Fieldbus™, Profibus®, Modbus, and other communication protocols for networking purposes.
6. They typically rely on a device placed in the flowstream to make a flow measurement; this device may interact with the flowstream and may cause pressure drop.

TABLE 3.2
New-Technology and DP Flowmeter Principles of Operation

Flowmeter Type	Technology
Coriolis	Fluid is passed through a vibrating tube; this causes the tube to twist. Mass flow is proportional to the amount of twisting by the tube
Magnetic	Creates a magnetic field within a pipe, typically using electrical coils. As electrically conductive fluid moves through the pipe, it generates a voltage. Flowrate is proportional to amount of voltage, which is detected by electrodes
Ultrasonic – single path transit time	Measures the time it takes for an ultrasonic pulse or wave to travel from one side of a pipe to the other. This time is proportional to flowrate
Ultrasonic – multipath transit time	Uses multiple ultrasonic paths to calculate flowrate; typical paths are three to eight
Ultrasonic – Doppler	Calculates flowrate based on the shift in frequency observed when ultrasonic waves bounce off particles in the flowstream
Vortex	A bluff body is placed in a flowstream; as flow passes this bluff body, vortices are generated. The flowmeter counts the number of vortices; flowrate is proportional to the frequency of vortices generated
Thermal	Heat is introduced into the flowstream and the flowmeter measures how much heat dissipates, using one or more temperature sensors
DP – orifice plate	A flat metal plate with an opening in it; a DP transmitter measures pressure drop and calculates flowrate
DP – Venturi tube	A flow tube with a tapered inlet and a diverging exit; a DP transmitter measures pressure drop and calculates flowrate
DP – pitot tube	An L-shaped tube inserted into a flowstream that measures impact and static pressure. The opening of the L-shaped tube faces directly into the flowstream. The difference between impact and static pressure is proportional to flowrate
DP – averaging Pitot tube	A Pitot tube having multiple ports to measure impact and static pressure at different points. Flowrate is calculated by DP transmitter based on average of difference in pressure readings at different points
DP – flow nozzle	A flow tube with a smooth entry and sharp exit; flowrate is calculated based on difference between upstream and downstream pressure

Conventional flowmeters include DP flowmeters (which include primary elements combined with transmitters), positive displacement, turbine, open channel, and VA flowmeters. Business is brisk with many of these meters. Why are customers still so loyal to these meters?

Familiarity Breeds Respect

While the explanations vary with the type of meter, there are several themes that run throughout. One answer is *familiarity*. End-users like having a technology they are familiar with and can understand. DP, positive displacement, and turbine meters especially are very well known and understood technologies. There is a comfort level among users with these technologies that is less likely to exist with the newer technologies such as Coriolis and vortex. In case more meters need to be added in a plant, users often stick with what they have rather than selecting a different type of meter.

A second reason is *installed base*. Some flowmeters such as DP and positive displacement have been around for over 100 years. Once these meters are installed, customers find in many cases that it is easier to replace them with meters of the same kind than to switch to another technology. Once a technology is in place, backup parts are readily available, any potential problems are usually known, and the path for replacement is clear. All these are reasons to stick with an existing technology.

Users are also sticking with conventional technologies because *suppliers are bringing out improved products*. Turbine suppliers are using material such as ceramic to improve the life of ball bearings. Rosemount has introduced the 3051S, a pressure transmitter with increased accuracy and stability. Positive displacement suppliers are using enhanced manufacturing techniques to build more precision into their positive displacement meters. Communication protocols such as HART and Profibus are beginning to appear on turbine and positive displacement meters. All these changes are resulting in improved and more reliable meters for users to choose from.

Switching Technologies Has a Cost

While end-users are not averse to changing technologies, they are not likely to do so unless they have a specific reason to make this change. One reason is having a problem with the flowmeter. Another is being bought out and having to go with the technology from a new company. Still another is budget requirements that dictate going to a less expensive meter. But changing technologies is not without cost. It usually means taking time to learn a new technology, finding a new supplier, and stocking a different set of backup parts. All these cost time and/or money.

Another reason why users continue to stay with the conventional technologies is that they are genuinely the best solution for certain types of flow applications. Each type of meter has its own set of applications in which it excels. This varies by meter type.

Differential Pressure

Differential pressure (DP) flowmeters are among the most conventional of flowmeters. Their large installed base means that they still exert a powerful force on the flowmeter market.

What Is a DP Flowmeter

A DP flowmeter consists of a DP transmitter integrated with a primary element and has the capability of calculating flowrate based on difference in pressure. DP flowmeters rely on a constriction placed in the flow line that creates reduced pressure in the line after the constriction. A primary element is used to create the constriction in the flowstream. A DP flowmeter also requires a means to detect the difference in upstream versus downstream pressure in the flow line. While this can be done with a manometer, today's DP flowmeters use DP transmitters that sense the difference in pressure, and then use this value to compute flowrate.

When most flowmeters are sold, the transmitter and sensor are sold together. This is true for ultrasonic, vortex, Coriolis, turbine, and other types of flowmeters. All these flowmeters operate based on a correlation between flowrate, or mass flow, and some physical phenomenon. For ultrasonic flowmeters, it's the difference in transit time of sound waves sent across the pipe. For turbine flowmeters, it's the speed of the rotor. DP flowmeters also correlate flow with a physical phenomenon; the difference in pressure upstream and downstream from a constriction in the flowstream.

Where DP flowmeters differ from other flowmeter types is that the transmitter is still often sold separately from the primary element that creates the constriction in the flowstream, sometimes from different suppliers. Because a DP flow transmitter cannot make a flow measurement without a primary element, customers who purchase a DP flow transmitter without the primary element are not actually buying a DP "flowmeter." They don't have a DP flowmeter until they connect the primary element to the DP flow transmitter. *Thus, a DP flowmeter is considered to be a DP flow transmitter that is connected to a primary element for the purpose of making a flow measurement.*

In the past, the pressure transmitter companies sold DP transmitters and users ordered their primary elements separately. Now, however, a number of companies are selling DP transmitters already integrated with a primary element, such as an Annubar or an orifice plate. It is tempting to consider these the only true DP flowmeters; however, a better description is that these are DP flowmeters with an integrated primary element. If a customer assembles a DP flowmeter by connecting up a DP flow transmitter to an orifice plate or a Venturi tube from another source, the result is just as much a DP flowmeter as an integrated product.

To Stay with DP or Switch

Many end-users of DP transmitters today are facing a dilemma. Should they upgrade to new-technology flowmeters, or stick with the tried and true DP meters? Of course, some end-users at large plants may choose to upgrade some of their DP meters and stick with DP technology elsewhere in the plant. Instead of replacing all their DP meters, they may make this decision on a case-by-case basis.

DP flowmeters are sometimes contrasted with new-technology meters because many users today are replacing their DP and other conventional flowmeters with ultrasonic, Coriolis, and other meters of recent vintage. New-technology meters have four traits in common: They were introduced after 1950, they incorporate recent technological advances, they are the focus of new product development, and they perform at a higher level than conventional flowmeters. New-technology meters include ultrasonic, Coriolis, magnetic, vortex, and thermal.

Where DP Meters Excel, and Considerations

Generally, DP meters excel at measuring clean liquids, steams, and gases when pressure drop is not a major issue, the application requires low to medium accuracy, and price is a consideration. However, certain designs and certain primary elements can handle dirtier fluids, or provide higher accuracy, or address other challenges. All DP flowmeters cause varying amounts of pressure drop, depending on the type of primary element used; however, some primary elements cause more, some less. For example, orifice plates cause substantial loss of line pressure, while averaging pitot tubes cause less. DP flowmeters are considerably less expensive to buy than most Coriolis and ultrasonic meters.

Paradigm Case Application

The paradigm case application for DP flowmeters is *clean flowing liquids, steams, and gases, where pressure drop is not a major concern and high accuracy is not required.* DP flowmeters consist of DP transmitters that have a primary element attached. The primary element places a constriction in the line to create a pressure differential. The DP transmitter measures this pressure difference and uses this information to compute flowrate.

Primary Elements

A primary element is something designed to create a constriction in the flowstream that creates reduced pressure in the line after the constriction. This is what DP flowmeters rely on, by detecting the difference in upstream versus downstream pressure in the flow line and then calculating the flowrate. The history of primary elements and DP flow measurement goes back to at least the 1700s.

The main types of primary elements include orifice measuring points, pitot tubes, Venturi tubes, cone elements, flow nozzles, wedge elements, and some others such as Dall tubes and laminar flow elements. Each of these designs has particular uses or limitations, different enough that there is no one "primary elements paradigm case" to cover all of them except in that all are used in DP flow measurement and would thus come under the DP paradigm case.

Positive Displacement

Positive displacement meters are highly accurate meters that are widely used for custody transfer applications. They are widely used for custody transfer of commercial and industrial water. They are also used for custody transfer of hydrocarbon liquids to and from delivery trucks. Positive displacement meters have the advantage that they have been approved by a number of regulatory bodies for this purpose, and they have not yet been displaced by other applications.

Paradigm Case Application

The paradigm case application for positive displacement meters is for *clean, noncorrosive liquids and gases that are flowing at a low flowrate or that are viscous.* Because positive displacement meters capture the flow in a container of known quantity, speed of flow doesn't matter to a positive displacement meter. They also do well with liquids that would give other meters headaches, like honey and other highly viscous fluids.

Turbine

Like positive displacement meters, turbine meters are used for custody transfer of commercial and industrial water and other liquids. They are also used for custody transfer of hydrocarbon-based liquids. And like positive displacement meters, turbine meters are widely used for custody transfer of natural gas.

Paradigm Case Application

The paradigm case application for turbine flowmeters is for *clean, steady, medium to high speed flowing liquids or gases.* While positive displacement meters excel at measuring low flows, turbine meters do best with steady, high-speed flows. Turbine meters also are more adaptable to large pipe sizes, including sizes over 12 inches, than are positive displacement meters. There are at least eight different types of turbine meters, and each one is designed for a specific set of applications.

Open Channel

Open channel meters are the only game in town when it comes to open channel applications. However, there are different types of open channel meters. Open channel applications are defined as those in which liquid flows in a stream or conduit that is not closed, or liquid flowing in partially full pipes that are not pressurized.

Some open channel flowmeters require hydraulic structures such as weirs or flumes. These are similar to the primary elements used with DP meters. Flow passes through the weir or flume, and flowrate is calculated based on the level or depth of the flow as it passes. Another popular method is called velocity area. Using this method, the velocity of the stream is computed by one method (e.g., electromagnetic), and the level or depth of the stream is computed, usually by another method (e.g., radar). These values are then used to calculate flowrate, although the area of the flow must also be known.

Paradigm Case Application

The paradigm case application for open channel flowmeters *depends on the type of open channel meter.*

Weirs: For open channel flowmeters using *weirs*, the paradigm case application is for *free-flowing streams with sufficient slope for relatively rapid flow through the weir.*

Flumes are normally used only when weirs will not work, so their paradigm case applications should be defined accordingly. The paradigm case for flumes is for *high velocity free flowing streams that contain some sediment or solids.*

Area velocity flowmeters' paradigm case is for *temporary flow measurements of flowing liquids in partially or completely filled pipes where the use of hydraulic structures is not practical.*

VARIABLE AREA

Most variable area (VA) flowmeters consist of a tapered tube that contains a float. The upward force of the fluid is counterbalanced by the force of gravity. The point at which the float stays constant indicates the volumetric flowrate, which can be often read on a scale on the meter tube. VA meter tubes are made of metal, glass, and plastic. Metal tubes are the most expensive type, while the plastic tubes are lower in cost. Metal tubes are used for high-pressure applications.

Paradigm Case Application

The paradigm case application for VA flowmeters is for *clean flowing liquids and gases where high accuracy is not required.* While most VA meters can be read manually, some also contain transmitters that generate an output signal that can be sent to a controller or recorder. While VA meters should not be selected when high accuracy is a requirement, they do very well when a visual indication of flow is sufficient. They are very effective at measuring low flowrates and can also serve as flow/no-flow indicators. VA meters do not require electric power and can safely be used in flammable environments.

SELECTING A FLOWMETER

Which flowmeter is best for a given application depends on the characteristics of the application. If the application includes very low flowrates, or viscous flow, positive displacement or thermal flowmeters are good choices to consider. For very high gas flowrates, a turbine or ultrasonic flowmeter is a likely choice. Every flowmeter has its own paradigm case applications, which include those in which it excels. The following tables review the paradigm case conditions for new-technology and conventional flowmeters. An application may fit the paradigm case for more than one flowmeter, although not always.

Surveys of flowmeter users consistently show that reliability and accuracy are the two performance criteria that are rated highest in importance by users when selecting flowmeters. Among new-technology flowmeters, Coriolis flowmeters provide the highest accuracy, followed by ultrasonic and magnetic meters. In terms of cost, many users are now distinguishing between purchase cost and cost of ownership. As a result, they may be willing to pay more for a flowmeter if it promises reduced maintenance costs.

How are decisions actually made in a plant about what flowmeter to buy? Often, users choose to replace like with like, for several reasons. Often, inventories of parts and supplies are built up in a plant based on a particular type of flowmeter. It can be very expensive to train personnel to install, use, and maintain a new type of flowmeter. Changing flowmeter types sometimes means changing flowmeter suppliers, which can be difficult.

The above helps to explain why DP flowmeters still have the largest installed base of any flowmeter type. The battle for the hearts and minds of users is largely between the suppliers of new-technology flowmeters and the suppliers of DP flowmeters. It is less of a battle among the suppliers of new-technology flowmeters, although new lower-cost Coriolis flowmeters may begin to impinge on the magmeter market.

Multivariable flowmeters represent one way that DP flowmeter suppliers are responding to the challenge of new-technology flowmeters. Multivariable flowmeters usually measure pressure and temperature, in addition to flow. Multivariable vortex and multivariable magnetic flowmeters have also been developed, and it is likely that more types of multivariable flowmeters will be introduced in the future. This ongoing drama is definitely worth watching.

REVIEW TABLES OF PARADIGM CASE CONDITIONS

Paradigm Case Selection Method

While various selection methods have been devised, this chapter presents a step-by-step method that begins by matching the application involved with the paradigm case applications for various types of flowmeters. It then advocates looking at application, performance, cost, and supplier criteria in order to select a flowmeter. A statement of this paradigm case method follows.

Table 3.3 sums up the paradigm case conditions for new-technology flowmeters. Table 3.4 sums up the paradigm case conditions for conventional flowmeters.

An application may fit the paradigm case for more than one flowmeter, although not always. Surveys of flowmeter users consistently show that reliability and accuracy are the two performance criteria that are rated highest in importance by users when selecting flowmeters. Among new-technology flowmeters, Coriolis flowmeters provide the highest accuracy, followed by ultrasonic and magnetic meters. In

TABLE 3.3
Paradigm Case Conditions for New-Technology Flowmeters

Flowmeter Type	Paradigm Case Conditions	Comment
Coriolis	Clean liquids and gases flowing sufficiently fast through pipes two inches or less in diameter	Measure mass flow directly; some provide density measurement. Can handle some fluids with varying densities that other meters types cannot. High initial cost but normally low maintenance costs; generally limited to smaller line sizes
Magnetic	Conductive liquids flowing through full pipes that do not contain materials that could damage the liner or coat the electrodes	Wide range of liner types; wide range of line sizes. Do not work with gas or steam; cannot measure nonconductive fluids
Ultrasonic – transit time	Clean, swirl-free liquids and gases of known profile	High accuracy may require the use of a multipath meter
Vortex	Clean, low viscosity, swirl-free, medium to high-speed fluids	Work for liquids, gas, and steam
Thermal	Clean, pure gases or mixtures of clean, pure gases whose composition is known, and whose heat capacities are also known	Widely used in measuring the flows of clean gases; can handle very low flows. Very limited application to liquids; do not work for steam; low to medium accuracy

terms of cost, many users are now distinguishing between purchase cost and cost of ownership. As a result, they may be willing to pay more for a flowmeter if it promises reduced maintenance costs.

Paradigm Case Selection Steps

1. Every type of flowmeter is based on a physical principle that correlates flow with some set of conditions. This principle determines the paradigm case application for this type of flowmeter. When selecting a flowmeter, begin by selecting the types of flowmeters whose *paradigm case applications* are closest to your own.
2. Make a list of *application criteria* that relate to the flow measurement you wish to make. These conditions may include type of fluid (liquid, steam, gas, slurry), type of measurement (volumetric or mass flow), pipe size, process pressure, process temperature, condition of fluid (clean or dirty), flow profile considerations, fluid viscosity, fluid density, Reynolds number constraints, range, and others. From those types of flowmeters selected in step 1, select those that best meet these application criteria.

TABLE 3.4
Paradigm Case Conditions for Conventional Flowmeters

Flowmeter Type	Paradigm Case Conditions	Comments
Differential pressure	Clean flowing liquids, steams, and gases, where pressure drop is not a major concern and high accuracy is not required	Lower cost; some newer technological improvements; low to medium accuracy; pressure drop; orifice plates subject to wear
Positive displacement	Clean, non-corrosive liquids and gases that are flowing at a low flowrate or that are viscous	Highly accurate; many approvals; widely used for custody transfer; can handle low flows and viscous flows; moving parts subject to wear
Turbine	Clean, steady, medium- to high-speed flowing liquids or gases	Adapts better to larger lines sizes than positive displacement; many approvals; widely used for custody transfer; bearings subject to wear; limited ability to handle impurities
Open channel	*Depends on the type:Weirs:* free-flowing streams with sufficient slope for relatively rapid flow through weir *Flumes:* high velocity free flowing streams that contain some sediment or solids *Area velocity flowmeters:* temporary flow measurements of flowing liquids in partially or completely filled pipes where use of hydraulic structures are not practical	For rivers, streams, partially filled large pipes; medium accuracy
Variable area	Clean flowing liquids and gases where high accuracy is not required	Visual flow indication; can serve as flow/no-flow indicators; can handle low flowrates; low accuracy; many without transmitters

3. Make a list of *performance criteria* that apply to the flowmeter you wish to select. These include reliability, accuracy, repeatability, range, and others. From those types of flowmeters selected in step 2, select the ones that meet these performance criteria.
4. Make a list of *cost criteria* that apply to your flowmeter selection. These include initial cost, cost of ownership, installation cost, maintenance cost, and others. From the types of flowmeters chosen in step 3, select the types that meet your cost conditions.
5. Make a list of *supplier criteria* that govern your selection of a flowmeter supplier. These include type of flowmeter, company location, service,

responsiveness, training, internal requirements, and others. From the types of flowmeters listed in step 4, select the suppliers that meet your criteria.

6. For the final step, *review* the meters that are left as a result of step 4 and the suppliers listed as a result of step 5. *Review* the application, performance, and cost conditions for the remaining flowmeter types, and *select* the one that best meets all these conditions. Now *select* the best supplier for this flowmeter from those suppliers listed as a result of step 5.

4 Coriolis Flowmeters

OVERVIEW

Coriolis flowmeters are the most accurate flowmeter made. While many magnetic flowmeters have accuracies in the range of 0.5 percent, many Coriolis flowmeters achieve accuracy of 0.1 or even 0.05 percent. The high accuracy of Coriolis flowmeters is one of the major reasons for the extremely rapid growth in their use over the past five years. Companies that need flowmeters for custody transfer, or want highly accurate measurements in terms of mass, have several good reasons to select Coriolis flowmeters.

Coriolis meters have a reputation for being high-priced meters. However, most users today take into account the distinction between purchase cost and cost of ownership, or life cycle cost. Even though Coriolis meters have a higher purchase cost than many other flowmeters, they may cost less over the lifetime of the meter due to reduced maintenance costs. Unlike turbine and positive displacement meters, Coriolis meters do not have any moving parts, apart from the vibrating tube. They are not subject to wear in the way that orifice plates are. With many companies reducing their engineering and maintenance staffs, having a meter that does not require a great deal of maintenance can be a major advantage.

Suppliers have made a number of improvements in Coriolis technology over the past five years. Coriolis meters are now much better able to measure gases than previously, and thus a majority of Coriolis suppliers now have meters that can measure gas flow. Straight tube meters have become more accurate and reliable, thereby addressing some of the drawbacks of bent tube meters. These include pressure drop, the ability to measure high-speed fluids, and the tendency of bent tubes to impede the progress of fluids. And Micro Motion and Endress+Hauser have broken price barriers in offering Coriolis meters for considerably a lower cost than they did before.

The large majority of Coriolis flowmeters have in the past been sold for line sizes below two inches. This is still true today. However, in the past several years, a number of companies have brought out Coriolis flowmeters in line sizes above six inches. Despite their higher cost, companies are using these meters due to their high accuracy and reliability. Companies that have entered the large line size market include Rheonik (now independent from GE Measurement), Endress+Hauser, Micro Motion, and Shanghai Yinuo KROHNE.

Up until 1994, nearly all Coriolis meters were bent tube meters. While bent tube Coriolis meters have advantages over many traditional technology meters, they do introduce pressure drop into the system.

DOI: 10.1201/9781003130017-4

TABLE 4.1
Advantages and Disadvantages of Coriolis Flowmeters

Advantages	Disadvantages
High accuracy	High initial cost
Approved for custody transfer	Becomes expensive and unwieldy in line sizes above four inches
Now available for line sizes above six inches	Gas flow measurement can be difficult due to low density of gas
Can handle sanitary applications	Pressure drop for bent tube meters
Excel in line sizes of two inches and less	Can have a problem measuring liquids with entrained gas
High reliability; low maintenance	
Much new product development ongoing	
Have custody transfer approvals for liquid and gas applications	

Endress+Hauser introduced the first straight-tube Coriolis meter in 1987. This meter had dual tubes and later evolved into the ProMass. Schlumberger brought out a straight tube flowmeter in 1993 but withdrew it after several months. KROHNE introduced the first commercially successful single straight tube Coriolis meter in 1994. Since that time, this type has become increasingly popular with Coriolis users.

Straight tube meters address the problem of pressure drop because the fluid does not have to travel around a bend. This makes them better able to handle high velocity fluids.

Straight tube meters can be drained more easily, which is important for sanitary applications. Liquids do not negotiate a bend or curve to build up and leave residue on.

Straight tube meters also have a more compact design. Bent tube meters can be quite large and unwieldy, especially in the larger sizes. Having a more compact meter can be an advantage where space is a concern (Table 4.1).

Coriolis flowmeters fill the need for a flowmeter that measures mass directly. Many other flowmeters measure velocity or volume; in case a mass flow measurement is needed, it is calculated, using pressure and temperature values. Many of these flowmeters are affected by temperature, pressure, viscosity, and density. By measuring mass flow directly as the fluid passes through the meter, Coriolis meters make a measurement that is virtually independent of changing process variables such as temperature and pressure. As a result, Coriolis meters can be used on a variety of fluids without recalibration and without the need to compensate for process variables that are specific to a particular kind of fluid.

A REVIEW OF EARLY CORIOLIS PATENTS

There were a number of patents filed in the 1950s and 1960s for Coriolis or Coriolis-like flowmeters. Even earlier, in 1917, Ernest F. Fisher on behalf of General Electric Company filed a patent called "Fluid Flow-Indicating Mechanism." As part of this patent, it is claimed:

> "It is an established principle that vibrations will be set up in fluid flowing past a statuary object and that the periodicity of the vibrations will be proportional to the velocity of flow of the fluid."

While this sounds somewhat like the formula for a vortex flowmeter, and it attempts to measure velocity rather than mass flow, it does introduce the idea of vibration of the fluid.

In a patent approved in July 1952, Paul Kollsman presented an "Apparatus for Measuring Weight Flow of Fluids." While it does not mention a Coriolis force, it does introduce the idea of measuring the weight or mass of a fluid. To quote from the patent:

> The present invention provides an apparatus for measuring the flow of fluids and provides, more particularly, an apparatus for determining the actual weight flow of a fluid which may be a gas, a liquid, or a mixture of both, the measurement being accurate within a wide range of density and viscosity of the fluid, and uninfluenced by the pressure of the fluid.

The following year, in January 1953, John M. Pearson, on behalf of the Sun Oil Company, patented a meter using a gyroscopic principle. This patent is simply called "FLOWMETER," and the meter was designed to measure mass flow.

This invention relates to a flowmeter designed to measure flow in terms of mass of the fluid.

While this patent does not mention the Coriolis force, it does describe the following mechanism:

> A curved conduit, means for leading fluid to and from said conduit, means for imparting to said conduit angular movement about an axis transverse to an axis about which there exists angular momentum of the flowing fluid through the conduit, and means, including a device sensitive to gyroscopic couples of said conduit transverse to the axis of movement of the conduit, for indicating mass flow through the conduit.

Interestingly, this patent cites no previous patent references.

A patent in April 1958, filed by Roby B. White on behalf of American Radiator & Standard Sanitary Corporation is called "CORIOLIS MASS FLOWMETER." This appears to be the earliest patent that mentions the "Coriolis force." The flowmeter is described as follows:

> The present invention relates to instruments for measuring the mass rate of flow of fluids and in particular to an improved flowmeter of the type in which mass flow rate is made responsive to Coriolis force ... In instruments of the class described the fluid to be measured is subjected to tangential acceleration in a whirling tube, or impeller, the torque exerted on the tube in reaction to the Coriolis force of the accelerated fluid being measured as an indication of the mass flow rate.

In May 1960, Yao Tzu Li patented an invention called "MASS FLOWMETER" that involves rotating the flow:

> The present invention operates by causing the fluid to be rotated as it flows radially outward from an axis. This produces a Coriolis acceleration in the fluid, and therefore a Coriolis force is applied by the fluid to the member through which the fluid flows. This force is measured, and the mass rate flow of the fluid is obtained.

Interestingly, Yao Tzu Li cites Fisher (1917), Kollsman (1952), and Pearson (1953) as references, but does not cite White (1958), who specifies a mechanism for rotating the fluid and also mentions the Coriolis force.

In November 1965, July 1967, and December 1967, Anatole J. Sipin patented three separate devices designed to measure mass flow. These designs by Sipin were designed to improve on earlier designs (described above) that involved rotating the flow or using an oscillating gyroscope. Sipin's objection to rotating gyroscopic designs was that they required rotating fluid seals, which introduced potential leakage and friction problems. His objection to oscillating gyroscopic designs was that they introduced bends and turns that made the meter difficult to clean. Meter size was also an issue. According to Sipin, "the diameter of the loop must be ten to twenty or more times as large as the diameter of the flow passage … (making) the apparatus large and cumbersome for the flow range." This problem of meter size is one that Coriolis manufacturers have never fully solved, except for meters two inches or less in diameter.

In August 1972, James E. Smith patented a "Balanced Mass-Moment Balance Beam with Electrically Conductive Pivots." Beginning in August 1978, James (Jim) Smith began patenting a series of devices that became the basis for the flowmeters produced by Micro Motion. This patent was filed in June 1975. These patents explicitly evoke the Coriolis force. Jim Smith founded Micro Motion out of his garage in 1977. In 1981, Micro Motion introduced its first single bent tube Coriolis flowmeter, although the company introduced its first Coriolis meter in 1977, designed for laboratory use (the A meter).

CORIOLIS FLOWMETER COMPANIES

Emerson – Micro Motion

Emerson is a global technology and engineering corporation that manufactures and sells innovative products and solutions to industrial, commercial, and consumer markets. Emerson designs and manufactures electronic and electrical equipment, software, systems, and services for industrial, commercial, and consumer markets worldwide through two main businesses: The Automation Solutions business provides gas and liquid flowmeters and other products that automate and optimize production, processing, and distribution facilities for process, hybrid, and discrete manufacturers. The Commercial & Residential Solutions business offers products, systems, and software to enhance productivity, efficiency, and compliance.

Micro Motion is a subsidiary of Emerson in the Automation Solutions division, and a sister company to Rosemount. Micro Motion offers Coriolis flow and density measurement technology, as well as complementary expertise and other solutions including online density, temperature, and viscosity measurement for multiple critical process applications, including custody transfer. Micro Motion is the worldwide market leader in Coriolis flowmeters and has maintained that rank for decades.

History

Micro Motion was founded in 1977 by Jim Smith. Micro Motion released the first Coriolis meter in 1977 (the A meter). The A meter was introduced as a laboratory instrument, and development quickly progressed through the B model and C model in 1981, when the company introduced its first single bent tube meter. In 1984, Emerson Electric acquired Micro Motion. Emerson Electric owns 100 percent of Micro Motion stock.

Micro Motion maintains its headquarters and a manufacturing location in Boulder, Colorado, and has additional manufacturing locations in The Netherlands, Mexico, China, and Japan. The company has approximately 100 direct salespeople throughout the United States.

In July 2020, Emerson announced it would invest more than $100 million in Boulder, Colorado – the home of Micro Motion – to significantly expand its manufacturing space and launch a new innovation center focused on research, new product development and industry training for its advanced flow measurement products.

Coriolis Flowmeters

Micro Motion's Coriolis flowmeters are based upon a range of bent and straight flow tubes in a range of product families that can operate within wide ranges of temperature and pressure. Micro Motion also offers specialty flow tubes to address specific applications. In general, Micro Motion products have a proven track record in the chemical, food and beverage, life sciences, gas, oil and gas, and refining industries, where high accuracy is a primary application requirement.

The Elite Series has gained significant traction in the market due to its versatility and high- performance parameters. Models within the series are capable of volumetric flow, mass flow, and density measurement of liquids, gases, and multiphase fluids.

Foxboro by Schneider Electric

Schneider Electric is an international company focused on the use of technologies to address the worldwide issues of energy management and automation. It manufactures an array of electrical power products ranging from sensors and circuit breakers to signaling devices, motor starters, and voltage transformers. The company's primary industrial markets consist of electrical utilities, oil and gas, water and waste treatment plants, public-sector infrastructure, and marine.

Schneider Electric took ownership of Foxboro Company when it acquired Foxboro's parent company, Invensys, in February 2014. Foxboro was a good fit within the Schneider structure as it is a manufacturer of quality process measurement instruments designed to function within sophisticated device networks. This enabled Schneider to expand its native product portfolio and thus the total value proposition it could offer new and existing customers.

Schneider Electric is present in more than 115 countries. Revenue sources are divided quite evenly among three regions (Western Europe at 26 percent, Asia/Pacific at 30 percent, and North America at 29 percent), with the rest of the world contributing the remaining 15 percent.

History and Organization

Schneider Electric has an extensive corporate history, having been founded in the first half of the 19th century by the Schneider brothers. The company soon became an early entrant into the newly established European electricity market.

Its early innovations led to an association with the American company Westinghouse, and soon Schneider Electric was manufacturing electric motors, electrical equipment for power stations, and electric locomotives. Schneider has continued its activity in the energy markets through today, and now maintains a vertically integrated leadership position in electricity distribution, from smart grid networking systems through industrial, commercial, and residential energy products and services.

Foxboro, founded in Foxboro, Massachusetts in 1908, has more than 100 years of experience in industries such as chemicals, oil and gas, power generation, food, paper, and mining. The Foxboro Company was acquired by Siebe plc of Windsor, United Kingdom in 1990. In 1999, Siebe merged with BTR plc, another large British engineering firm, to form Invensys plc. Schneider Electric acquired Invensys, including Foxboro, in January 2014. In addition to a variety of other products and product support services, Foxboro by Schneider Electric manufactures and supplies Coriolis, magnetic, vortex flowmeters, and differential pressure transmitters and primary element assemblies to a worldwide marketplace.

Foxboro by Schneider Electric Coriolis Flowmeters

The company's Coriolis flowmeter line is designed for use with liquids, gases, and steam. Foxboro offers a variety of transmitters, which can be packaged with specific sensors to better meet the most exacting application requirements.

The CFT51 transmitter is an enhanced version of the original CFT50, which was the first transmitter to support two-phase performance. It is designed to measure and provide the mass flow and density of liquid, gas, and steam and does not require gas to be removed from the liquid stream or for batch processes to start or finish empty. Foxboro claims that the CFT51's 16 millisecond response time is the fastest on the market today. The CFT51 can be combined with a wide choice of mass flow tubes to provide mass, density, and temperature measurement and can be combined with any Foxboro CFS Series mass flow tube to form an I/A Series mass flow and density flowmeter.

CORIOLIS FLOWMETER THEORY OF OPERATION

While early Coriolis flowmeter patents relied on rotating the fluid or on rotating gyroscopes, these patents do not appear to have resulted in actual flowmeter products. It was the work of Jim Smith, founder of Micro Motion, to turn the Coriolis force patents into real products. Since that time, a host of companies have manufactured Coriolis meters, many of whom have their own patents. Some of these companies are covered in this chapter and are well known today, like Endress +Hauser, KROHNE, Yokogawa, Siemens, and Foxboro. Others have either been acquired or failed in their attempt to produce Coriolis meters. Examples include EXAC Corporation and Direct Measurement Corporation, though there are others. EXAC Corporation was absorbed by Emerson Micro Motion. Direct Measurement Corporation (DMC) was acquired by FMC Technologies but was unsuccessful in launching its straight tube gas flow Coriolis meter.

Modern Coriolis flowmeters today typically consist of one or two vibrating tubes with an inlet and an outlet. These tubes can either be straight or bent, though the large majority are bent. Whether they are single or dual, bent or straight, Coriolis flowmeters rely on oscillating tubes. The tubes are made to oscillate at their natural resonant frequency by an electromagnetic exciter or drive coil located at the apex of the tubes. The apex is the highest point of the tube, and it is where the inlet ends, and the outlet begins. Another way of describing the apex is that it is the central point between the beginning of the tube and the end of the tube.

The peak amplitude of vibration is at the apex of the flow tube or tubes. Magnet and coil assemblies called pickoffs are mounted at the same corresponding place on the inlet and the outlet portions of the flow tube(s). As the tubes oscillate, the voltage generated from each pickoff creates a sine wave. When there is no flow, the inlet and outlet sine waves are in phase. Being in phase means that they are in a synchronized motion.

Stop the Presses!

It is easy to hear words, and the words make sense, but often we think we understand how something works because all the words strung together make sense, and we can repeat them word for word, but yet may not completely grasp them. For example, you probably understand what condensation is – you see it happen every day when your glasses fog up, water forms on your cold iced tea glass, or when clouds form in the atmosphere. Nearly everyone knows that condensation happens when air becomes too heavy with water and so the water condenses out into liquid form.

Unfortunately, some of the most common experiences we have depend on things we cannot see. We cannot see atoms and molecules, so we can't really see how condensation occurs; we can only see that it occurs. When water vapor condenses to form liquid, the gas molecules slow down, come together, and form a liquid. This tends to happen when liquid becomes colder, and it is how clouds form. The opposite process, evaporation, can be explained in a similar way. Heat breaks up the liquid molecules and causes them to escape into space.

Two closely related examples are the formation of snow and rain. Few weather-related phenomena are as familiar as snow and rain, depending on the climate, of course. While both rain and snow are very familiar to most people, explaining how they occur might not be so easy for many people. Both involve complex processes including evaporation, condensation, heating and cooling, and gravity. Figure 4.1 is a graphic showing how snow forms.

FIGURE 4.1 How snow forms.

Figure 4.2 is a similar graphic showing how rain forms. The main difference between precipitation falling as snow and precipitation falling as rain is the temperature of the clouds aloft – and the temperature on the ground. Of course, we sometimes experience variations on this theme, like sleet, freezing rain, or even hail.

Some flowmeters play an important role in weather forecasting. Anemometers are used to measure wind speed, and as we shall see in Chapter 8, thermal flowmeters were developed out of anemometers. Doppler radar is used to measure the speed of objects such as rain drops and what direction they are moving. Doppler radar measures the shift in phase between a transmitted pulse and a received echo. Doppler radar works very much like ultrasonic Doppler flowmeters. And as we shall see very shortly, Coriolis flowmeters are named after Gustav Coriolis, who observed the effect of the rotation of the earth on weather systems near the equator.

Looking into the Soul of a Coriolis Flowmeter

We are about to stare into the soul of a Coriolis flowmeter, so let's make sure we understand the terms. First, we say the tube oscillates, or if there are two, the tubes oscillate. This means they go back and forth. Oscillation is a to and fro motion, like a pendulum. It is not rotation; it is oscillation. As explained earlier, the Coriolis tubes are made to oscillate by an electromagnetic exciter.

Magnet and coil assemblies are mounted at corresponding locations on the inlet and outlet portions of the tubes. If you are familiar with the theory behind magnetic flowmeters, fluid flowing through a magnetic field creates a voltage that is proportional to flowrate. The voltage is detected by electrodes and sent to the transmitter, where flowrate is calculated.

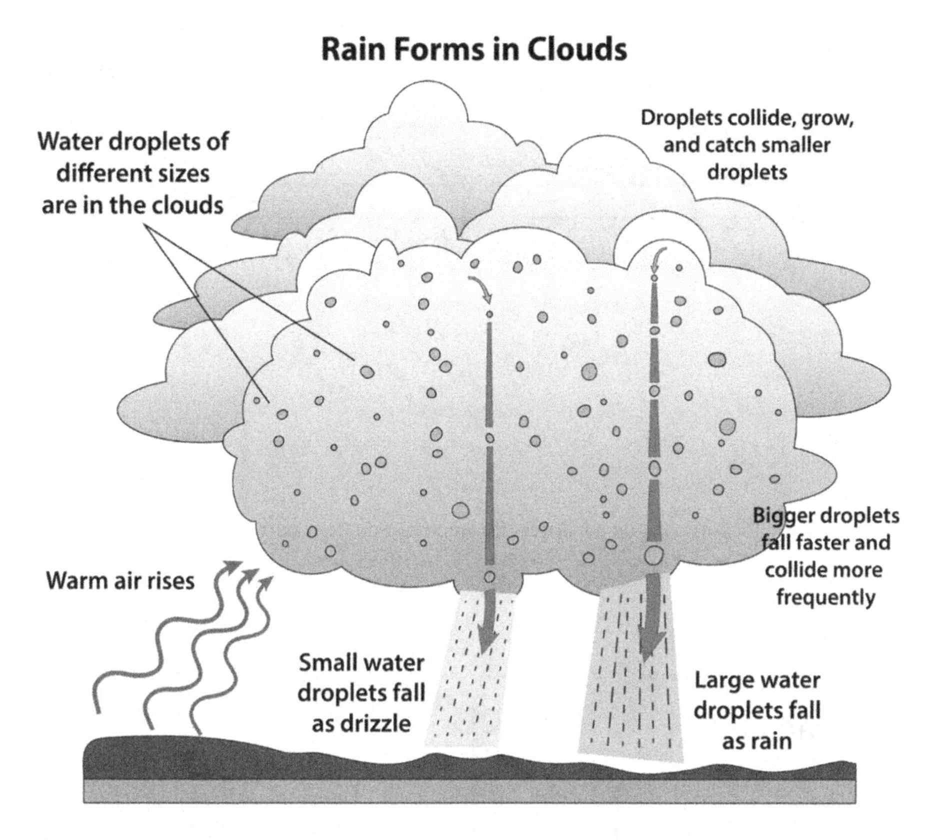

FIGURE 4.2 How rain forms.

In the case of Coriolis flowmeters, the meter has a pickoff coil on the inlet side and the outlet side of the flow tube. The pickoff contains a coil and a magnet. As the coil moves through the magnetic field from the vibration of the tubes, voltage is produced. This voltage can be represented as a sine wave. The exciter and the pickoff coil are arranged as in the following illustration of a straight tube meter (Figure 4.3).

In the following diagram, 2 and 4 are the pickoff coils. As they move through the magnetic field created by the magnet and pickoff coil, their motion can be represented as a sine wave.

Figure 4.4 shows the exciter that causes the tube to oscillate. The pickoff coils (here called pickup sensors) are located on either side of the apex of the tube. As the tubes oscillate, the voltage created from each pickoff generates a sine wave. These sine waves are analyzed by the transmitter.

When fluid is not moving through the tube, the inlet and outlet sine waves are in phase. Being in phase means that they are in a synchronized motion. This means that the waves are moving at the same rate and exactly together. When two people synchronize their watches, they set them to the same time so that the watches are moving together, and both tell the same time. Under no flow conditions, the sine

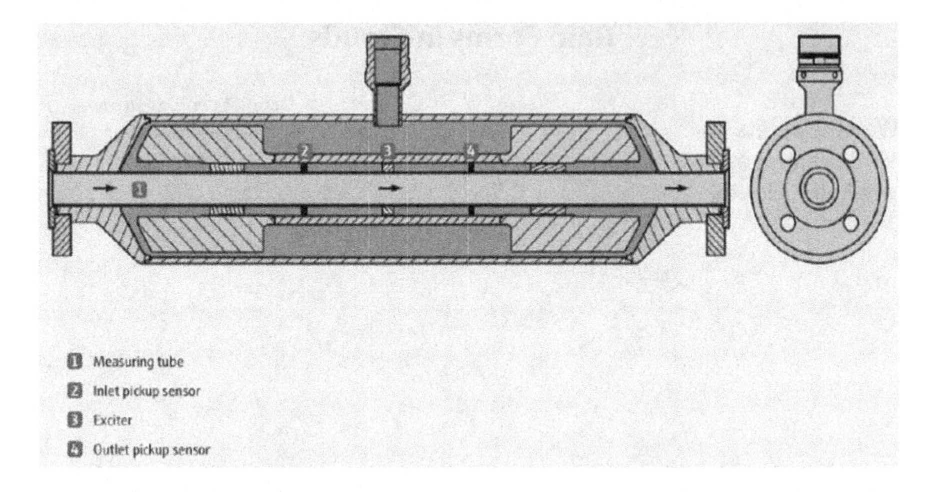

FIGURE 4.3 Pickoff coils on a Coriolis flowmeter. (Source: Courtesy of Endress+Hauser.)

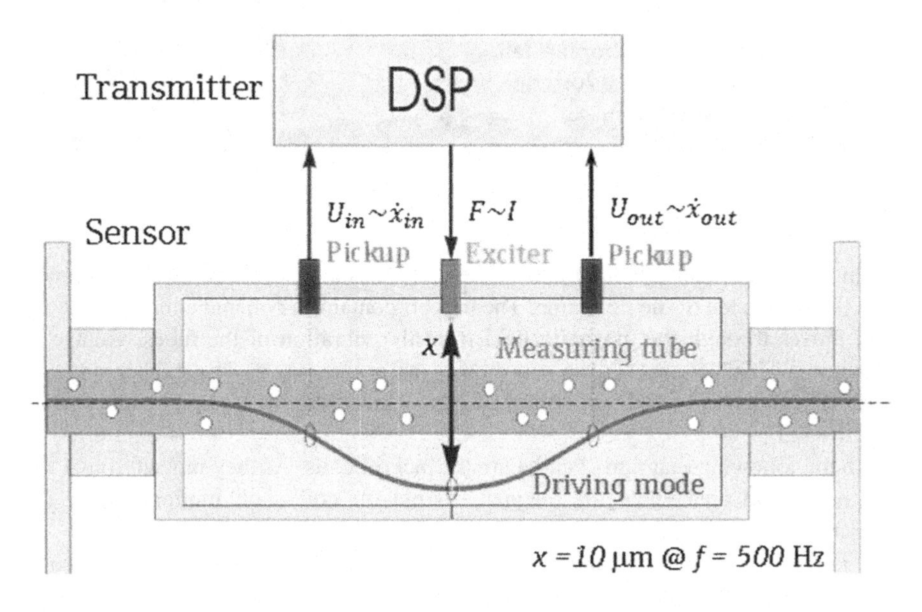

FIGURE 4.4 The exciter that causes a Coriolis tube to oscillate. (Source: Courtesy of Endress+Hauser.)

waves generated by the pickoffs on the inlet and outlet side look exactly the same (Figure 4.5).

When fluid moves through the tube, the inertial force of the fluid causes the tube to oscillate. This results in a phase shift, or time difference, between the sine waves

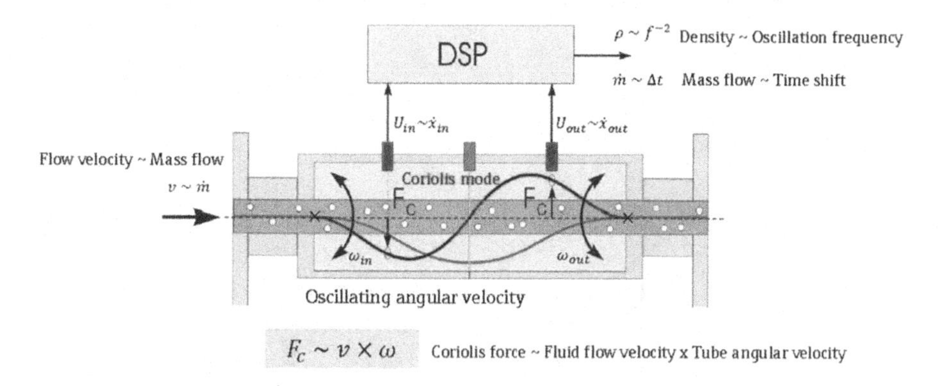

FIGURE 4.5 A Coriolis flowmeter with a twisting tube as a result of the inertial force of the fluid, resulting in a phase shift. (Source: Courtesy of Endress+Hauser.)

on the inlet side of the tube and the sine waves on the outlet side of the tube. The sine waves generated by the pickoff coils on the inlet and outlet side of the tube are no longer in phase; instead, they are asynchronous. There is now a difference in time between these sine waves, which is measured in microseconds. This difference in time is called delta t (Figure 4.6). Delta t is directly proportional to mass flowrate. The mass flowrate is computed by the transmitter, which outputs this value along with other desired values such as density, volumetric flow, and temperature.

While the amount of the phase shift or delta t is directly proportional to mass flow, the sine wave frequency indicates density. Frequency means the number of waves per second. A heavy fluid like honey will have a lower frequency than a lighter liquid such as water. Some Coriolis meters are used to measure density rather than flow, but generally both values are desired.

Coriolis Effect

Where does the Coriolis force come in? Coriolis flowmeters get their name from a French mathematician named Gaspard Gustave de Coriolis. In 1835, he wrote a paper describing how objects behave in a rotating frame of reference. Coriolis studied the transfer of energy in rotating systems like water wheels. When some people talk about how the Coriolis principle works, they give the example of the Coriolis effect. The Coriolis effect is not the result of a force acting directly on an object, but rather the perceived motion of a body moving in a straight line over a rotating body or frame of reference.

One common example given to illustrate the Coriolis effect is that of a ball propelled through the air a long distance in a straight line from the North Pole toward a target on the equator (see the Coriolis effect diagram in Figure 4.7). By the time the ball arrives at the equator, it will not land at its apparent target because the earth will have rotated sufficiently underneath the moving ball so that it will land some distance away from the perceived target on the equator. From the perspective of the person standing where the ball is “thrown,” the ball will appear to have curved.

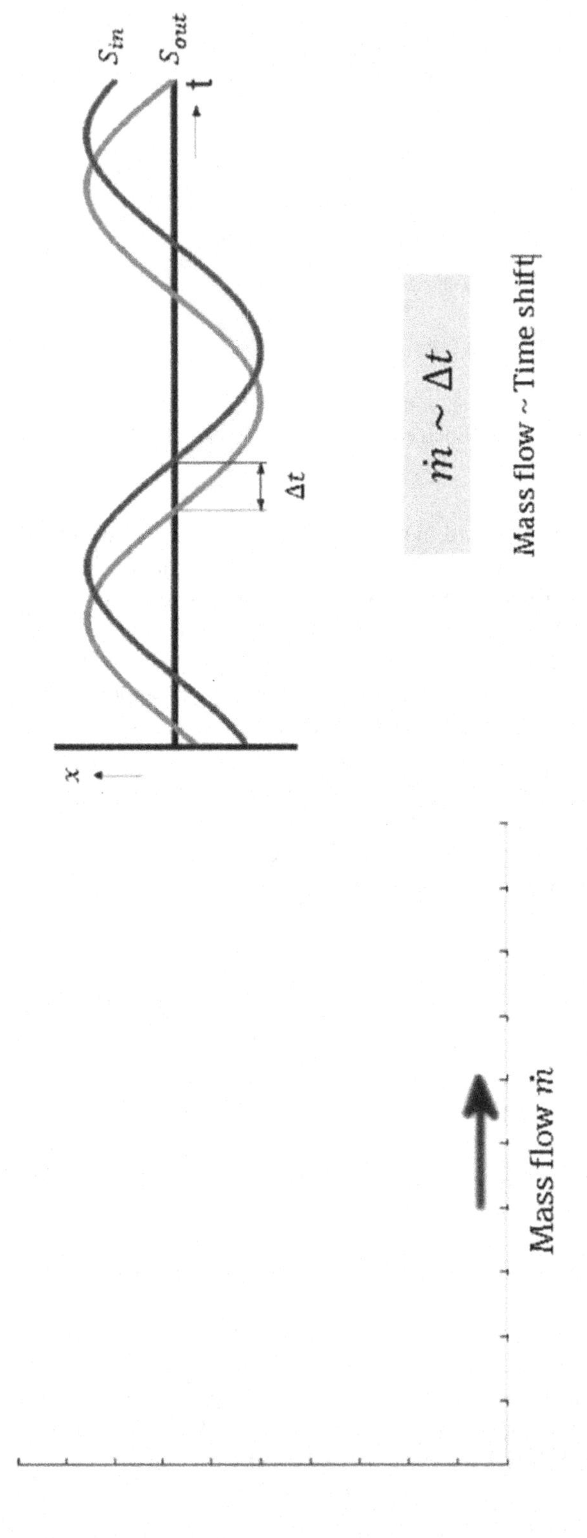

FIGURE 4.6 The delta t between sine waves that is proportional to mass flow. (Source: Courtesy of Endress+Hauser.)

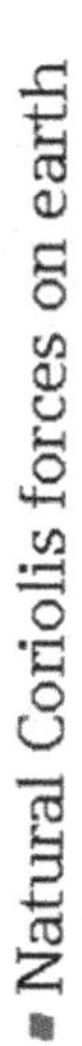

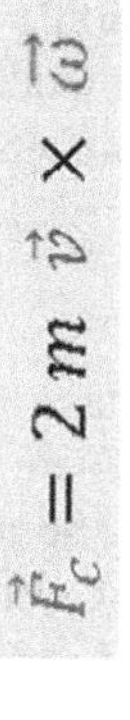

m Mass of moving object

$\vec{v}$ Radial linear velocity of object

$\vec{\omega}$ Angular velocity of earth

$\vec{F}_c$ Coriolis force acting on object

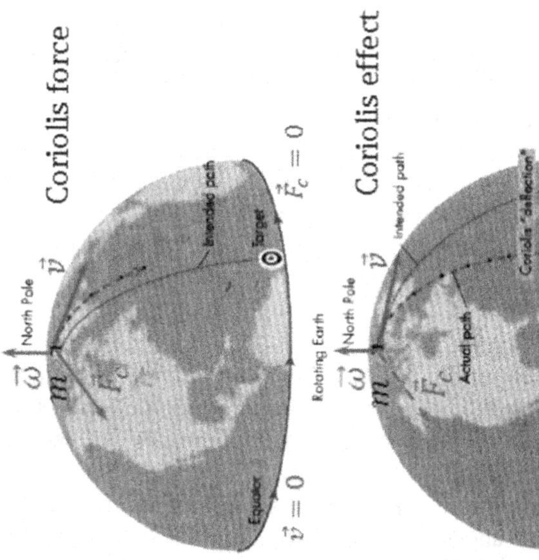

Coriolis force

- Weather systems
- Ocean currents
- River banks
- ...

Coriolis effect

- Railway tracks
- Foucaults pendulum
- ...

FIGURE 4.7 Natural Coriolis force on earth and the Coriolis effect as seen from the earth's rotation.

A similar example is someone throwing a ball from the center of a rotating merry-go-round. If she throws the ball at a horse on the edge of the merry-go-round, assuming the merry-go-round is rotating at a rapid pace, she will not hit the horse she is aiming at but perhaps the one after it, since the horse she aimed at will have moved by the time the ball reaches the edge of the merry-go-round.

In both these cases, there is no force acting to push the ball in a curved direction. Instead, its motion will appear to be curved from the perspective of the ball-thrower because the frame of reference is moving underneath the ball. This is why the Coriolis "force" is more appropriately called the Coriolis effect. It refers to the apparent effect on the motion of an object passing over a rotating frame of reference when viewed from the perspective of the point of origin of that moving object.

The Coriolis effect is also discussed in meteorological examples. Because the earth is constantly rotating, it provides a rotating frame of reference. High pressure systems that move north from the equator will appear to veer to the right. This apparent deflection will be more pronounced the farther north they go. The reason is that on a sphere such as the earth, a point moving around the equator will move faster than a point moving farther north, such as in Greenland. A point in Greenland will be moving more slowly than a point on the equator because it has less distance to travel in the same amount of time. This causes the apparent deflection of a weather system moving north, or a cannonball shot a long way north from the equator.

Coriolis Force

Some people describe the Coriolis effect as a "fictitious" force because it only involves apparent motion based on an observer having a particular point of view. This applies when an object is passing over a rotating frame of reference, as in the examples given above.

In reviewing the patents on early types of Coriolis meters, and also the Jim Smith patents, the inventors spoke about a Coriolis force, not a Coriolis effect. And in the description above, it is the inertial force of the fluid exerted on the oscillating tubes that causes the tube to twist, generating the phase difference between the inlet and outlet portions of the tube. This phase difference is directly proportional to mass flow.

The top half of the earth in the above diagram is meant to illustrate the effect of rotation on objects that are a part of the frame of reference, not objects that are flying over a moving frame of reference. For example, the earth's rotation together with the gravitational pull of the sun and moon create the earth's tides. The earth's rotation is responsible for the existence of day and night. It also affects deep ocean waves.

How Coriolis Flowmeters Came to Be

It is not completely clear how we came from Gustave Coriolis' analysis of the rotation of water wheels to Coriolis flowmeters. There does seem to be some confusion between Coriolis force and Coriolis effect. Some people who attempt to explain how Coriolis flowmeters work appeal to the Coriolis effect as an analogy. Yet there seems to be little relation between the Coriolis effect, which is acknowledged to be "fictitious," and the principle of operation of Coriolis flowmeters.

FIGURE 4.8 A Coriolis flowmeter mounted inline in a European calibration facility.

One possible explanation is that the early inventors designed instruments that actually rotated the fluid, and so they called them Coriolis flowmeters. This makes a connection between rotational motion and the workings of Coriolis meters. Later inventors abandoned the idea of rotating the fluid in favor of oscillating tubes. But because their patents cited the earlier patents that appealed to fluid rotation, they kept the name "Coriolis" to describe their meters.

The remaining question is whether the twisting motion of an oscillating tube is a form of rotation, or if this can be described in the terminology of Coriolis rotation. The key variable, as shown in the equation in Figure 4.7 diagram, is "angular velocity." If this is a Coriolis force, then Coriolis meters are aptly named. If angular velocity can be accounted for by fluid inertia, then perhaps they should instead be called inertial mass meters. Alternatively, if the back and forth twisting motion of Coriolis meters is a form of rotation, they could also be called "rotational" meters. Whatever we call these meters, they still remain the most accurate flowmeter made (Figure 4.8).

GROWTH FACTORS FOR THE CORIOLIS FLOWMETER MARKET

Growth factors for Coriolis flowmeters are as follows:

- Custody transfer of natural gas is a promising application.
- Suppliers introduce lower cost meters.
- New product features improve the performance of Coriolis meters.
- Coriolis flowmeters have reduced maintenance requirements.

- Straight-tube meters eliminate some performance issues with bent-tube meters.
- Large-diameter meters open up new applications.

Custody Transfer of Natural Gas Is a Promising Application

Custody transfer of natural gas is a fast-growing market, especially with the increased popularity of natural gas as an energy source. Natural gas changes hands, or ownership, at a number of points between the producer and the end-user. These transfers occur at custody transfer points and are tightly regulated by standards groups such as the American Gas Association (AGA). Other geographic regions have their own regulatory bodies, such as the Canadian Energy regulator in Canada.

One important function of the AGA and the American Petroleum Institute (API) is to lay down standards or criteria for sellers and buyers to follow when transferring ownership of natural gas and petroleum liquids from one party to another. In the past, these groups have published reports on the use of orifice plate meters and turbine meters for use in the custody transfer of natural gas. The importance of these reports is illustrated by the example of ultrasonic flowmeters. In the mid-1990s, a European association of natural gas producers called Groupe Européen de Recherches Gazières (GERG) issued a report laying out criteria to govern the use of ultrasonic flowmeters in the custody transfer of natural gas. This resulted in a substantial boost in the sales of ultrasonic flowmeters for this purpose in Europe. In June 1998, the AGA published AGA-9, which also gave criteria for using ultrasonic flowmeters in natural gas custody transfer situations. This caused a substantial boost in the sales of these meters for that purpose, especially in the United States. The market for using ultrasonic meters to measure natural gas for custody transfer is now one of the fastest growing segments of the flowmeter market. The AGA approved a report on the use of Coriolis flowmeters for custody transfer of natural gas in 2003. This report is called AGA-11. This report partially explains the overall positive growth rate of Coriolis flowmeters, as operators now use them for natural gas custody transfer applications. Even though it can often take some time for end-users to adopt a new technology, this report has provided a significant boost to the use of Coriolis flowmeters for natural gas flow measurement.

The API issued a draft standard entitled Measurement of Single-Phase, Intermediate, and Finished Hydrocarbon Fluids by Coriolis Meters. This document was added to the API Library in July 2012. A second draft standard called Measurement of Crude Oil by Coriolis Meters has also been approved by the API. And in 2013, a revised and complete API standard was published that describes methods for achieving custody transfer levels of accuracy when a Coriolis meter is used to measure liquid hydrocarbons. Topics covered included applicable API standards used in the operation of Coriolis meters, proving and verification using both mass- and volume-based methods, and installation, operation, and maintenance requirements. Both mass- and volume-based calculation procedures for proving and quantity determination are also included.

This approval of compliance with regulatory measurement standards has opened the door for Coriolis flowmeters to participate at the highest level within the liquid and gas hydrocarbon fluid measurement market.

Suppliers Introduce Lower Cost Meters

Micro Motion has introduced a series of Coriolis meters called the R-Series. These meters have broken a price barrier and are available in the $4,000 range. This addresses one of the major stumbling blocks to the use of Coriolis meters: high initial purchase price. It makes Coriolis meters more competitive with magnetic, vortex, ultrasonic, and positive displacement meters. While these other meters are typically available for somewhat less than Coriolis meters, even the R-Series meters, it substantially narrows the gap between Coriolis and these other meters in terms of initial purchase cost. Plus, the long-term benefit of comparatively less maintenance remains.

The reduced cost does come at a price, however. The R-Series meters are not as accurate as other Coriolis meters, such as those in Micro Motion's Elite Series. Published mass flow accuracies are in the range of ±0.5 percent of flowrate, which makes them comparable to many vortex flowmeters in terms of accuracy. Users who are looking for repeatability as much as accuracy may find that these meters satisfy their needs. They will also appeal to users who have always wanted a Coriolis meter, but who felt they were simply too expensive to buy.

Endress+Hauser, another quality mainstream supplier, has also released a low-cost Coriolis meter. It is also likely that other low-cost Coriolis meters will come on the market. Low-cost meters represent an important breakthrough for Coriolis suppliers in their attempt to gain further market penetration for Coriolis meters.

New Product Features Improve the Performance of Coriolis Meters

Suppliers have made a number of improvements in Coriolis technology over the past five years. Coriolis meters are now much better able to measure gases than previously. The majority of Coriolis suppliers now have meters that can measure gas flow. Straight tube meters have become more accurate and reliable, thereby addressing some of the drawbacks of bent tube meters. These include pressure drop, the ability to measure high-speed fluids, and tendency of bent tubes to impede the progress of fluids. Other improvements include the use of stronger and lighter materials of construction like titanium that make the meters more rugged and durable. Other sensor tube materials such as super duplex increase the level of pitting corrosion resistance, making the flowmeter capable of handling corrosive fluids that previously would have distorted flowrate measurements due to the internal erosion of sensor tubes made from of stainless steel.

Coriolis Flowmeters Have Reduced Maintenance Requirements

Coriolis meters have a reputation for being high-priced meters. However, most users today take into account the distinction between purchase cost and cost of ownership, or lifecycle cost. Even though Coriolis meters have a higher purchase price than many other flowmeters, they may cost less over the lifetime of the meter due to reduced maintenance costs. Unlike turbine and positive displacement meters, Coriolis meters do not have any moving parts apart from the vibrating tube. They are not subject to wear in the way that orifice plates are. With many companies reducing their engineering and maintenance staffs, or having their staffs reduced through retirement, having a meter that does not require a great deal of maintenance can be a major advantage, with savings accruing over time.

Straight-Tube Meters Eliminate Some Performance Issues with Bent Tube Meters

While bent tube meters still have advantages over many conventional meters, they do introduce pressure drop into the system. Pressure drop is an issue because in many cases the fluid has to be speeded up back to its original velocity. This costs money, as it requires the use of pumps. Another issue has to do with the tendency for build-up to occur around pipe curvatures. This can be a special problem for sanitary applications. Having a bent pipe also slows down the fluid, making it more difficult to meter high-velocity fluids.

Endress+Hauser introduced the first straight-tube meter in 1987. This meter had dual tubes and later evolved into the Promass. KROHNE introduced the first commercially viable single straight tube Coriolis meter in 1994. Since that time, this type has become increasingly popular with Coriolis users. Straight tube meters address the problem of pressure drop because the fluid does not have to travel around a bend. This makes them better able to handle high velocity fluids. Straight tube meters can be drained more easily, which is important for sanitary applications. Liquids do not have a bend or curve to build up and leave residue on. Straight tube meters also have a more compact design. Bent tube meters can be quite large and unwieldy, especially in the larger sizes. Having a more compact meter can be an advantage where space is a concern.

Straight tube meters are quite popular in the pharmaceutical industry. However, not all pharmaceutical applications are appropriate for straight tube meters. Most straight tube meters use titanium, which can leach into certain products. As a result, stainless steel is used for some of these applications where this type of leaching can occur. In many chemical applications, pressure drop is not an issue, so both straight and bent tube meters are used. Straight tube meters are used in the chemical industry for measuring polymers, which have high viscosities. Straight tube meters are also widely used in the food processing industry because they can easily be cleaned.

Large Diameter Meters Open Up New Applications

More than any other meter, Coriolis meters have line size limitations. Due to the nature of the technology, Coriolis meters get large and unwieldy in line sizes above four inches. Even two-inch, three-inch, and four-inch meters are quite large. The majority of Coriolis meters are in the zero to two-inch diameter ranges.

How will the larger size meters grow? One way is exemplified by Rheonik. Rheonik put together two six-inch Coriolis meters to create a meter that can handle larger line sizes. Perhaps lighter materials or other technological advances will make it possible to create more manageable Coriolis meters with large diameters.

Such innovation was used in making straight tube meters. Any progress in this area is likely to be welcomed by customers who would like to have smaller and more compact Coriolis flowmeters with diameters above two inches.

Other companies that have introduced Coriolis flowmeters for line sizes above six inches include Endress+Hauser, KROHNE, Micro Motion, and Shanghai Yinuo. Endress+Hauser and Micro Motion have bent tube meters, while KROHNE's large-size meters are straight tube. While the straight tube meters are long, they are less bulky than the bent tube meters of the corresponding size. There is definitely a trend among Coriolis suppliers toward offering flowmeters for the larger line sizes. Currently, there are Coriolis meters available in 8-, 10-, 12-, and 14-inch line sizes, and Endress+Hauser has even introduced a meter that will accommodate line sizes of 16 inches. Most of these meters are aimed at the custody transfer market for oil and gas applications.

Endress+Hauser's large line size offering is the Proline Promass X500, which is among the largest Coriolis flowmeters made, and is suited for use with nominal pipeline diameters ranging from 12 to 16 inches. The design produces a measurement error of ±0.10 percent on a standard basis, but up to ±0.05 percent on an optional basis. What is unique about the design of the Promass X500 is that it is a four-tube Coriolis meter and it is suitable for custody transfer applications.

Indications are that these large line size meters are selling quite well. Why are they doing so well, considering that their price tags are generally greater than $50,000? The answer has to do with the enhanced value of the fluids they are measuring. Many of these meters are designed to go into the oil and gas industry. While the price of oil declined significantly during the COVID-19 pandemic, it has come up from its lowest point and early in 2022 it was selling above $100 per barrel. Oil exploration and production activity is expected to pick up in 2022 and 2023, despite the long-term move to renewable energy. The enhanced accuracy provided by Coriolis flowmeters can make a significant difference in the measurement of high-priced fluids. Coriolis meters also offer high reliability and minimal maintenance, so their long-term cost of ownership is less despite their high initial cost. Look for more large line size Coriolis flowmeters to be released as suppliers continue to do research and development on this market.

FRONTIERS OF RESEARCH

Large Line Size Coriolis Meters

Building large and larger line size Coriolis flowmeters is a frontier of research for this meter type. The largest line size currently built is a 16-inch meter build by Endress+Hauser. It is not clear why a larger Coriolis meter cannot be built, but such a meter would have to overcome several barriers:

1. The meter would presumably have to be even larger and heavier than the current large line size meters. This would make it even more difficult to move around and install than existing meters. So far no one has been able to build a 20-inch meter.

2. As the meter gets larger, it becomes more difficult to vibrate the meter in such a way that it can reliably indicate mass flow. This is due to the increased weight of the meter.
3. Any meter that is larger than the existing meters would be even more expensive than the existing line size meters, some of which sell for $75,000. End-users would likely look for an alternative technology such as ultrasonic or turbine if these other meters could satisfy their needs for considerably less cost.

There once was a company in Israel that claimed to be working on a 24-inch Coriolis meter. However, this company has closed down in the past five years and can no longer be contacted.

One possible way this problem can be approached is by using ever lighter materials of construction for the flow tubes. This would facilitate the twisting motion and also make the meter lighter.

Explaining the Coriolis Force or Effect

There is a problem with the theory behind Coriolis flowmeters. Coriolis meters operate on the momentum of the fluid and rely on this momentum to deflect the vibrating metering tube. When fluid moves through the tube, the inertial force of the fluid causes the tube to twist. This results in a phase shift, or time difference, between the sine waves on the inlet side of the tube and the sine waves on the outlet side of the tube. This phase shift is known as delta t, and it is directly proportional to mass flow.

This account of the theory of Coriolis meters has nothing to do with the so-called Coriolis effect or Coriolis force. The term "Coriolis force" first appeared in the early 1900s as a result of the observations of Gustave Coriolis in the mid-1800s about the effect of the earth's rotation on large weather systems near the equator. Somehow this Coriolis force or effect was translated into the result of throwing a ball from the North Pole or Greenland toward the equator and having it miss its target due to the earth's rotation. A similar idea is advanced about throwing a ball to someone riding a horse on a merry-go-round. The idea appears to make sense because the theory is about the behavior of objects in a rotating plane. However, it is difficult to view a vibrating meter tube as one that is rotating.

This idea may have come about because some early patents going back to the 1950s actually called for a meter with a rotating element. It may be that the term "Coriolis" came into use as applied to this meter at that point, and the name stuck. However, I do not believe that the meter patented by Jim Smith in the early 1970s actually rotated.

I have argued this point at length with the best minds at Micro Motion, KROHNE, and Endress+Hauser, both in Europe and in the United States. Regardless of the truth of the matter, it would be helpful to end-users and others in the field if someone could come up with a clearer explanation of the relation between Coriolis theory and the Coriolis force or effect.

Measuring Gas Flow

The percentage of Coriolis meters that measure gas flow still remains at a relatively low level. This is because liquid is denser than gas and it can be a problem for gas to cause sufficient twisting in the tubes to get the same kind of accurate measurement that liquids can create. This is especially a problem for straight tube meters. Gas flow measurement by Coriolis meters is a matter of ongoing research by the major suppliers of Coriolis meters.

5 Magnetic Flowmeters

Magnetic flowmeters have been around longer than any other new-technology flowmeters. The Tobinmeter Company first introduced magnetic flowmeters for commercial use in Holland in 1952. Foxboro introduced them to the United States in 1954. Since this time, more than 60 suppliers worldwide have come to offer magnetic flowmeters for sale.

Magnetic flowmeters generate more revenues worldwide than any other type of flowmeter. Revenues from magnetic flowmeters exceed revenues from all other flowmeter types, including Coriolis, positive displacement, turbine, and differential pressure (DP) meters. The story is somewhat different in terms of units, however. More DP and variable area flowmeters are sold annually than magnetic flowmeters. Despite this, the higher average selling price of magnetic flowmeters enables them to generate more revenues annually than these other types of meters.

Most flowmeters do their best work in clean liquids or gases. This is true, for example, of turbine, Coriolis, ultrasonic transit time, and vortex meters. Magnetic flowmeters, by contrast, thrive on dirty liquids. Magnetic flowmeters and Doppler ultrasonic meters are the only two of the main types of meters that do well in dirty and impure liquids, although DP meters can also measure dirty liquids if they have the right kind of primary element. Magnetic flowmeters are used to measure the flow of conductive liquids and slurries, including pulp and paper slurries and black liquor. Their main limitation is that they cannot measure hydrocarbons (which are nonconductive), and hence they are not widely used in the petroleum industry.

Magmeters, as they are often called, are highly accurate and do not create pressure drop. Their initial purchase cost is medium to high, depending on size. While their price is generally higher than DP flowmeters, most are priced lower than equivalent Coriolis meters.

WHY ARE MAGMETERS SO POPULAR

In addition to a large installed base, magmeters have many advantages that help account for their role as the leading revenue-generating flowmeter:

Willing to Do the Dirty Work

Most flowmeters do their best work in clean liquids or gases. Magnetic flowmeters, by contrast, thrive on dirty liquids. They are used to measure the flow of conductive liquids and slurries, including pulp and paper slurries and black liquor. Liners are the "secret sauce" of magnetic flowmeters, enabling them to measure both very dirty and very clean liquids. They can measure the dirty and caustic liquids and slurries common to the pulp and paper and wastewater industries, as well as the

DOI: 10.1201/9781003130017-5

hygienic and sanitary liquids common to the food and beverage and pharmaceutical industries.

Flexible

In addition to liner flexibility, magmeters are available in a wide range of sizes from less than 1/8 inch to over 100 inches. Furthermore, the development of insertion meters gives more options to end users who want to measure liquids in large line sizes at lower costs.

Accurate and Cost Effective

Magnetic flowmeters are highly accurate, do not create pressure drop, and can be used for custody transfer applications. Magnetic flowmeters do not have moving parts, and provide a highly stable measurement. Their initial purchase cost is medium to high, depending on size. While their price is generally higher than DP flowmeters, most are priced lower than equivalent Coriolis meters. In addition, advanced diagnostics are making magmeters both more intelligent and more reliable (Table 5.1).

TOP INDUSTRIAL USES

Magmeters are widely used in the water and wastewater industry. Thanks to new industry group standards that include using magmeters for water utility measurement, magmeters are now displacing positive displacement and turbine meters in some residential and industrial applications.

TABLE 5.1
Advantages and Disadvantages of Coriolis Flowmeters

Advantages	Disadvantages
Medium accuracy	Cannot be used to measure nonconductive fluids or hydrocarbon-based liquids
Approved for custody transfer for some water applications	Cannot measure gas or steam flow
Can handle sanitary applications	Medium to high initial cost, depending on line size
Can be used in large line sizes	Electrodes subject to coating
High reliability	Insertion meters have limited accuracy
Can handle measurement of dirty liquids	
Insertion meters available for large line sizes	
Many types of liners accommodate a wide variety of fluid types	
Do not create pressure drop	

Magnetic flowmeters are also widely used in the chemical, food and beverage, and pharmaceutical industries, due in part to the variety in lining choices that enable them to measure a wide variety of liquids. Our data shows an increase in the use of magmeters in these industries.

Magmeters are also making inroads in the oil and gas industry. Magmeters' main limitation is that they cannot measure hydrocarbons (which are nonconductive), and hence they have not been widely used in the petroleum industry. However, they have come to be widely used in hydraulic fracturing to measure the water injected into oil and gas wells for "fracking" as well as the water flowing from them for capture, disposal, or recycling.

ALTERNATING CURRENT VERSUS DIRECT CURRENT

Magnetic flowmeters use wire coils mounted onto or outside a pipe. A voltage is then applied to these coils, generating a magnetic field inside the pipe. As the conductive liquid passes through the pipe, a voltage is generated and detected by electrodes mounted on either side of the pipe. The flowmeter uses this value to compute the flowrate.

When magnetic flowmeters were first introduced, many had coils powered by continuous alternating current (AC). These devices had the disadvantage that they were subject to noise that interfered with the proper reading of the meter. As a result, they needed to be calibrated regularly against an onsite hydraulic zero to maintain their accuracy.

Direct current (DC) magmeters were developed to solve the problems from the noise associated with AC meters. The DC meters are based on pulsed direct current. When the current is turned on, a voltage is generated in the magnetic flowmeter, showing the velocity of a flowing liquid. When the current is turned off, any remaining voltage is assumed to be due to noise. The meter computes flow velocity by subtracting this extra remaining voltage.

While DC pulsed technology was first introduced in 1974, it became popular in the 1980s, and its popularity has grown since then. Many pulsed DC magmeters have the drawback, however, of lower signal strength. This gives AC meters an advantage for measuring some dirty liquids and slurries.

To compensate for low signal strength, some DC meter suppliers developed "high strength" DC meters. These high strength meters still use the pulsed on-off technology of DC meters, but they have a higher coil current. This makes them better able to handle high noise applications – such as slurries and dirty liquids – than standard DC meters. These high strength meters are growing rapidly in popularity.

LINERS – THE "SECRET SAUCE" OF MAGNETIC FLOWMETERS

Liners enable magmeters to measure both very dirty and very clean liquids. With the appropriate liner option, they can measure the dirty liquids and slurries common to the pulp and paper and wastewater industries, as well as the hygienic and sanitary liquids common to the food and beverage and pharmaceutical industries.

Of the nine main types of liners for magnetic flowmeters the most dominant are PFA (perfluoroalkoxy), PTFE (polytetrafluoroethylene) – Teflon®, a familiar trade name for PTFE made by DuPont – and hard rubber. Hard rubber is widely used for water and wastewater applications.

Liners increase magnetic flowmeters' durability and reliability, and make it possible to use them with almost any type of liquid. No other flowmeter that measures liquids has such versatility when it comes to the material in the flowmeter that makes contact with the liquid.

GREATER CONDUCTIVITY

The inability of magnetic flowmeters to measure nonconductive liquids will always be a barrier to their use in the oil and gas and refining industries, barring some unforeseen technological breakthrough. However, suppliers have succeeded in reducing the amount of conductivity required to measure flow, in part by boosting the amount of power used to excite the magnetic coils, thereby creating a stronger signal. By pushing back the boundaries of conductivity, suppliers are making magnetic flowmeters usable in a broader range of applications.

AREAS OF GROWTH FOR MAGNETIC FLOWMETERS

One application where magnetic flowmeters have shown the ability to replace other meters is in water utility measurement. This is an area that has long been dominated by positive displacement and turbine meters. One reason is that these meter types have long had industry approvals that magnetic flowmeters were lacking. As industry groups develop standards for the use of magnetic flowmeters in utility measurement, they are displacing positive displacement and turbine meters for some applications. These applications include both industrial and residential applications. While this book does not include residential meters, it does include the use of magnetic flowmeters for industrial applications. Both are potential areas of future growth for magnetic flowmeters.

Another area where magnetic flowmeters have shown growth over the past five years is the chemical industry. Magnetic flowmeters are widely used in the chemical, food and beverage, and pharmaceutical industries, due in part to the variety in choice of linings that make them able to measure a wide variety of liquids. Our data shows an increase in the use of magnetic flowmeters in the chemical, food and beverage, and pharmaceutical industries over the past three years. However, they also have to compete with Coriolis flowmeters, which typically offer higher accuracy. Straight-tube Coriolis meters are popular in these industries because the fluid does not build up around bent tubes and interfere with measurement accuracy.

The sheer size of the magnetic flowmeter market may also limit its growth rate. While revenues from magnetic flowmeter sales exceed those of any other type of meter, it may be that magnetic flowmeters have already penetrated the liquid flow measurement market to a great extent, and that future growth will come from the same industries and applications they have traditionally come from.

If the future growth of this market continues to come from traditional industries and applications, it will depend on growth in the water and wastewater industry, on in-plant measurement, and on replacement orders. Certainly, all these areas are likely to show growth in the future, due to population expansion and the increasing scarcity of clean and potable water. There will also be an increased need for wastewater treatment plants. And fracking will continue to be a growth area, especially as it spreads beyond the United States. Nonetheless, it is unlikely that the already large magnetic flowmeter market will grow at the same rate as the Coriolis or ultrasonic flowmeter markets. These latter two markets have a wider variety of applications than magnetic flowmeters and are also displacing other meter types at a faster rate.

IMPORTANT FACTS ABOUT MAGNETIC FLOWMETERS

Magnetic flowmeters can only be used to measure liquid flows; they cannot measure the flow of gas or steam.

The growth rate of the magnetic flowmeter market is not as great as that of Coriolis, but the magnetic flowmeter market is currently the largest of any single flowmeter type in terms of revenues. This is true despite the fact that magnetic flowmeters can only measure conductive liquids. This means they cannot measure steam, gas, or hydrocarbon liquids. However, suppliers have made some progress in enabling magnetic flowmeters to measure low conductivity liquids.

Magnetic flowmeters are the meter of choice for many liquid applications. Magmeters can easily handle slurries and other wastewater and dirty water applications, as many have liners that tolerate these types of liquids.

There is a wide variety of linings used for sanitary applications. PFA and PTFE are the two most popular types of liners. PTFE is a combination of fluorine and carbon that is more popularly known as Teflon. Other popular types of liners are hard rubber, ceramic, and polyurethane.

Magnetic flowmeters are also used for filling machines that dispense soda and other consumer-oriented beverages.

There is new growth in the magmeter market attributable to the hydrofracking applications being developed in oil and gas production.

A significant percentage of magnetic flowmeters come in sizes of one inch or less, and they can measure very low flows.

The most popular industry for magnetic flowmeters remains the water and wastewater industry, representing roughly 25 percent of worldwide sales revenues. Magmeters excel at measuring any kind of water, whether clean or dirty, and can measure slurries that few other flowmeters can tolerate. For clean water, they are relatively inexpensive, although inline magnetic flowmeters can fit line sizes of 120 inches or more. These meters are quite expensive. As a result, a number of suppliers have introduced insertion meters that are not as accurate but are significantly less expensive than their inline counterparts.

One main advantage of magnetic flowmeters is that they can be used in almost any line size from less than ½ inch to over 120 inches. They are among the most adaptable flowmeters for any line size. Large line size magmeters are used for large pipes in the water and wastewater industry, while small ones are used in food and pharmaceutical applications. In addition to the inline meters, magmeters also come in the form of insertion meters for large line sizes. While insertion magmeters are not as accurate as the inline versions, they are considerably less expensive.

NEW DEVELOPMENTS

While the magnetic flowmeter market is a mature and stable one, some new product developments favor continued growth.

One recent development is the advent and increasing popularity of two-wire magmeters. Four-wire meters have a dedicated power supply. Two-wire meters use the power available from the loop-power supply, reducing wiring and installation costs. While two-wire meters still represent only a small percentage of the total magnetic flowmeters sold, their use continues to grow.

Another important development is battery-operated and wireless magnetic flowmeters. Battery-operated meters make it possible to install magmeters in hard-to-reach places. And wireless meters can transmit a receivable signal where the use of wires is impractical. Both of these segments represent fast-growing areas of the magnetic flowmeter market.

Another recent development in product types is the trend toward lower-cost, compact magnetic meters in the United States. Compact meters have traditionally outsold remote meters in Europe, while the reverse has been true in the United States. While compact meters may be losing some ground to remote meters in Europe, they still dominate that market.

MAGNETIC FLOWMETER COMPANIES

ENDRESS+HAUSER

Endress+Hauser is a leading global supplier of products, solutions, and services for industrial process measurement and automation. The company manufactures and supplies products for mass and volumetric flow, level, pressure, tank gauging, temperature, and industrial liquid analytical instrumentation, as well as process recording equipment. The Endress+Hauser group encompasses 136 companies in 47 countries.

In flow measurement alone, the company is a leading supplier of magnetic flowmeters worldwide and also supplies Coriolis, DP, vortex, thermal mass, and ultrasonic flowmeters. Endress+Hauser is also a leading supplier of level devices to the European market.

History and Organization

George Endress and Ludwig Hauser founded Endress+Hauser in 1953 to specialize in level measurement. Since that time, the company has expanded to include all the major parameters in industrial measurement. Since 1975, the group has been wholly owned by the Endress family. Klaus Endress, a son of the founder, became chief executive officer (CEO) in 1995. Matthias Altendorf assumed responsibility for the day-to-day operation of the business in 2014 as the new CEO.

In early 2014, the company opened new offices in Abu Dhabi and Dubai, and a new sales center in Algeria to begin operating directly in the Mideast and North Africa. Heavy investments in oil and gas, and in developing public infrastructure such as saltwater desalination plants, water storage, and power stations, have opened up a strong market for high-grade measurement engineering and automation solutions. These offices helped the company take greater advantage of its strengths in these growth areas by providing local access to product services, engineering, and project management. In March 2020, the group established Endress+Hauser Middle East to further strengthen its presence on the Arabian Peninsula. The new organization, headquartered in Dubai, leads and supports all regional solutions, sales, and service activities, including Endress+Hauser sales centers and sales representatives in the Middle East region.

The Endress Group has a presence in 125 countries, with manufacturing plants in Brazil, China, Czech Republic, France, Germany, India, Italy, Japan, South Africa, Switzerland, the United Kingdom, and the United States. One-third of the group's employees work in the tri-border area of Switzerland, Germany, and France.

Magnetic Flowmeter Products

Endress+Hauser has sold two million magnetic flowmeters since 1977. In recent years, the company has consistently been among the top suppliers and remains one of the most innovative magmeter manufacturers. Endress+Hauser Proline magmeters are used in applications for water, acids, alkalis, slurries, and other conductive liquids, including monitoring, filling, dosing, and custody transfer. The company's magmeters, need no maintenance and integrate seamlessly into processes. They offer variety of transmitters and a wide range of lining types. Some models feature ultra-compact transmitters ideal for skid builders, equipment manufacturers and system integrators. Endress+Hauser's signature Heartbeat Technology ensures measurement reliability and compliant verification.

The robust Promag S is suitable for inhomogeneous, abrasive and corrosive fluids, including challenging applications in sewage treatment, pulp and paper production, or in the primaries and metals industry. In addition to measuring flow, Promag 55S also calculates solids content. The Proline Promag P high-temperature series is designed for chemical and process applications, including corrosive liquids and highest medium temperatures. Proline Promag W is designed for water and wastewater.

KROHNE

KROHNE is a global manufacturer and supplier of industrial process instrumentation, measurement solutions and services. The company offers its products and services across an industrial spectrum that includes oil and gas, water and wastewater, chemicals and petrochemical, power, mining, and many more industries. The company has unique expertise in flow, level, temperature, pressure measurement, and analysis instrumentation. KROHNE supplies magnetic, ultrasonic, variable area, vortex, and Coriolis flowmeters in addition to a portfolio of level, temperature, pressure, and analytical devices.

KROHNE has 16 production facilities in 11 countries, 49 KROHNE-owned companies and joint ventures, and 55 exclusive representatives worldwide. KROHNE provides application knowledge and local contacts for instrumentation projects in more than 100 countries.

History and Organization

The origins of KROHNE go back to 1921, when Ludwig Krohne began to produce variable area flowmeters. In 1936, the company moved from rented to company-owned buildings. The buildings, Ludwig, and his son were all lost to World War II, but his wife, Anna, kept the company running. In 1949, she brought Ludwig Krohne's grandson, Kristian Dubbick, on board. He oversaw KROHNE's growth from a small company with a workforce of eight people into a large flow measurement company.

In 1961, KROHNE engineers developed the first commercially available magnetic flowmeter for industrial use. In 1979, Kristian Dubbick resigned as managing director and became chairman of the board of directors. A limited partnership was formed that became the basis for the company today.

After almost 40 years as chairman of the advisory board of the KROHNE Group, Prof. Dr. Rolf Theenhaus retired on January 1, 2020, and passed the reins to Michael Rademacher-Dubbick, who had served for more than 25 years as managing shareholder.

As of February 2020, KROHNE Messtechnik GmbH has a new three-person management team. Dr. Michael Deilmann and Lars Lemke have been appointed as new managing directors and lead the company together with Ingo Wald. Dr. Deilmann is responsible for sensor development and mechanical production. Lemke is responsible for communications technology, integrated systems, and electronics production. Wald has been managing director of KROHNE Messtechnik since 2006 and is responsible for sales, finance, purchasing, IT, and human resources. Wald is also member of the KROHNE Group Executive Board as chief financial officer.

Magnetic Flowmeters

KROHNE magnetic flowmeter designs and liner materials adapt to a wide range of applications with conductive liquids, from basic applications and potable water to custody transfer and extremely adhesive, abrasive or aggressive fluids (Figure 5.1).

FIGURE 5.1 Three large diameter magnetic flowmeters await delivery. (Source: Courtesy of KROHNE.)

The OPTIFLUX series, perhaps the best-known member of KROHNE's magnetic flowmeter portfolio, is suitable for most industrial process control measurement applications. Models within the series include a unique diagnostics package that can look into a process, and a quick start function for simpler start-up. Other series are focused on more discrete application sets, including partially filled pipes, water and wastewater and agriculture industries, food and beverage and pharmaceutical industries, power and nuclear power industries, and single-use biopharmaceutical applications. In March 2021, KROHNE introduced the AF-E 400 ultra-compact magnetic flowmeter for utilities and industrial automation applications. It claims to be best-in-class in terms of temperature range, accuracy, pressure drop, and flow range. KROHNE also offers two insertion magmeters: OPTIPROBE for water and wastewater applications and DWM 2000 for liquids, pastes, and slurries, including immersed applications.

HOW THEY WORK

Magnetic flowmeters use Faraday's law of electromagnetic induction. According to this principle, a voltage is generated in a conductive medium when it passes through a magnetic field. This voltage is directly proportional to the density of the magnetic field, the length of the conductor, and the velocity of the conductive medium. In Faraday's law, these three values are multiplied together, along with a constant, to yield the magnitude of the voltage.

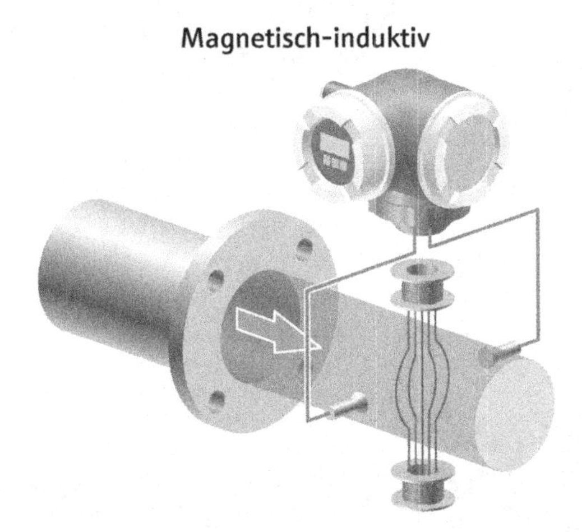

Magnetic flowmeter operating principle. (Source: Courtesy of Enfress+Hauser.)

Magnetic flowmeters use wire coils mounted onto or outside of a pipe. A voltage is then applied to these coils, generating a magnetic field inside the pipe. As the conductive liquid passes through the pipe, a voltage is generated and detected by electrodes, which are mounted on either side of the pipe. The flowmeter uses this value to compute the flowrate (Figure 5.2).

Magnetic flowmeters are used to measure the flow of conductive liquids and slurries, including paper pulp slurries and black liquor. Their main limitation is that they cannot measure hydrocarbons (which are nonconductive), and hence are not widely used in the petroleum industry. "Magmeters," as they are often called, are highly accurate and do not create pressure drop. Their initial purchase cost is relatively high, though most magmeters are priced lower than equivalent Coriolis meters.

GROWTH FACTORS FOR THE MAGNETIC FLOWMETER MARKET

- The installed base of magnetic flowmeters.
- New developments in two-wire and battery-operated magmeters.
- Companies are improving magnetic flowmeter technology.
- Magnetic flowmeters are very popular in Europe.
- Foundation Fieldbus and Profibus products are becoming more widely available.
- The large number of magnetic flowmeter suppliers.

FIGURE 5.2 A magnetic flowmeter. (Source: Courtesy Endress+Hauser.)

The Installed Base of Magnetic Flowmeters

One major growth factor for magnetic flowmeters is their large installed base worldwide. Magnetic flowmeters were first introduced for commercial use in 1952. This puts their time of introduction well before that of other new technology flowmeters, including Coriolis (1977), ultrasonic (1963), and vortex (1969). Because they were introduced so much earlier than other new-technology meters, magnetic flowmeters have had more time to penetrate the markets in Europe, North America, and Asia.

Located within the large installed base are industry segments that magnetic flowmeters diversely populate. While the water and wastewater industry remains the single largest industry segment for magmeters, the chemical and food and beverage industries together are much larger. This sort of industry diversity can expose magmeters to the benefits of growth within a single industry in which it may participate on a concentrated basis (e.g., chemicals over the last few years). It may also work against magmeters as well, such as in the delays to water and wastewater projects attributable to state or national governments dependent upon oil receipts, when receipts declined during the oil price depression of 2014–2016. These macroeconomic factors have played a large role in the growth profile of magnetic flowmeters during the last several years.

Installed base is a relevant growth factor because often when ordering flowmeters, especially for replacement purposes, users replace like with like. The investment in a flowmeter technology is more than just the cost of the meter itself. It also includes the time and money invested in training people how to install and use the meter. In addition, some companies stock spare parts or even spare meters for replacement purposes. As a result, when companies consider switching from one flowmeter technology to another, they typically consider more than just the purchase price. The large installed base of magnetic flowmeters worldwide will continue to be a source of orders for new meters in the future.

New Developments in Two-Wire and Battery-Operated Magmeters

While the magnetic flowmeter market is a mature and stable one, there are some new product developments in the magmeter market, and these new developments favor continued growth. One recent development is the advent of two-wire magnetic flowmeters. Four-wire meters have a dedicated power supply. Two-wire meters use the power available from the loop-power supply. This reduces wiring costs, and can result in lower installation costs. These meters are becoming more popular with users, due to the cost savings involved. While two-wire meters still represent only a small percentage of the total magnetic flowmeters sold, their use continues to grow.

Another important development is growth in battery-operated and wireless magnetic flowmeters. Battery-operated meters make it possible to install magmeters in hard-to-reach places. And wireless meters can transmit a receivable signal where the use of wires is impractical. Both of these segments represent fast-growing areas of the magnetic flowmeter market.

Suppliers have regularly brought out magnetic flowmeters with new liner types.

Companies Are Improving Magnetic Flowmeter Technology

The inability of magnetic flowmeters to measure nonconductive liquids will always be a barrier to their use in the oil and gas and refining industries, barring some unforeseen technological breakthrough. However, suppliers have succeeded in reducing the amount of conductivity required to measure flow with a magnetic flowmeter. This has been done in part by boosting the amount of power used to excite the magnetic coils, thereby creating a stronger signal. By pushing back the boundaries of conductivity, suppliers are making magnetic flowmeters usable in a broader range of applications.

In this respect, magnetic flowmeters are somewhat like transit time ultrasonic flowmeters. For many years, transit time meters could only be used on clean liquids. Then advanced signal processing technology made it possible to use transit time meters on liquids with some impurities. This has greatly increased the flexibility of transit time meters, and made them usable in a wider variety of applications. There remain, however, a group of more difficult applications where transit time meters cannot be used, and Doppler meters are required. These are applications with fluids that have significant amounts of impurities or solids. Likewise, there is a group of applications that are better handled with AC magnetic flowmeters than by DC magnetic meters. This group is very similar to the group of applications that are better handled by Doppler meters: applications for liquids that have significant amounts of impurities or solids.

Magnetic Flowmeters Are Very Popular in Europe

Magnetic flowmeters are especially popular in Europe. Magnetic flowmeters were first introduced in Holland in 1952. Water is a highly valued resource in Europe, and magmeters are widely used to measure the flow of water. Food processing and pulp and paper are both very prevalent industries in Europe, and magmeters are heavily used in both of these industries.

Europeans also seem to show a preference for spoolpiece over clamp-on meters, and most magnetic flowmeters are of the inline type, whether wafer or flanged. There are no clamp-on magnetic meters, but there are clamp-on ultrasonic meters, and ultrasonic meters are an alternative to magmeters for some applications. The leading supplier of one major alternative to magnetic flowmeters, DP flowmeters, is based in the United States. By contrast, the top two magnetic flowmeter suppliers are based in Europe. All these reasons are contributing factors to explaining the popularity of magnetic flowmeters in Europe.

Foundation Fieldbus and Profibus Products Are Becoming More Widely Available

After many years of discussion and debate, Foundation Fieldbus flowmeters are available for purchase. In order to be considered a Foundation Fieldbus flowmeter, a meter has to be approved by the Fieldbus Foundation as one that conforms to Foundation Fieldbus standards.

Foundation Fieldbus in particular has taken much longer than many people projected to become a reality. Most people who are buying replacement meters will not choose Foundation Fieldbus meters, since they are designed for networked environments. However, users who are building new plants, or who are renovating existing plants, are likely to go with the latest technology. These users may select Foundation Fieldbus or Profibus flowmeters to install in their new or renovated plants. The use of both Foundation Fieldbus and Profibus products is expected to increase as the installed base of both types of meters increases in size.

The penetration of the magnetic flowmeter market by Foundation Fieldbus and Profibus has been slower for magnetic flowmeters than for some other new-technology meters. This is because many magnetic flowmeters are sold into the water and wastewater industry, which has been slow to embrace fieldbus standards.

The Large Number of Magnetic Flowmeter Suppliers

There are more magnetic flowmeter suppliers than any other type of new technology meter except for ultrasonic meters. With over 60 suppliers worldwide, users are assured of having a steady stream of new products and product upgrades. Because the suppliers are distributed throughout the world, users can select high quality magnetic flowmeters from companies in their geographic region. The major European suppliers include Endress+Hauser, KROHNE, and ABB. Major suppliers in the United States include Emerson Rosemount and Foxboro by Schneider Electric. ABB also has manufacturing operations in the United States. In Japan, major suppliers include Yokogawa, azbil, and Toshiba.

Besides the major suppliers, there are a number of smaller suppliers that have specialty or niche products, often distributed mainly in their own geographic region. Europe has quite a few specialty suppliers of this type. Some of these companies only manufacture magmeters, while others manufacture several types of flowmeters. Still others manufacture magnetic flowmeters, but also resell other types of meters to broaden their product line. The presence of so many magnetic flowmeter suppliers worldwide will continue to be a growth factor for magmeters.

FRONTIERS OF RESEARCH

Measuring Liquids with Lower Conductivity Values

Magnetic flowmeter suppliers have cut into the market share of Doppler meters by improving the ability of magnetic flowmeters to measure liquids with lower conductivity values. Part of this success has been due to more advanced signal processing methods and more sophisticated software. This is still an area for ongoing research, and it can be expected that suppliers will continue to be able to measure ever low-conductivity liquids. No one has yet made a magnetic flowmeter that will measure the flow of hydrocarbons, however.

Continued Advances in Liners

Liners are the “secret sauce” in magnetic flowmeters. A wide variety of liners is already available. Liners are especially valuable for sanitary applications, and also for wastewater applications. Despite all the progress that has already been made, expect new types of liners as magnetic flowmeters expand their reach and new applications emerge.

A Continued Focus on Market Strengths

Magnetic flowmeters have strong penetration in their areas of strength: liquid measurement in the water and wastewater, chemical, food and beverage, pharmaceutical, and biotech industries. They can handle almost any line size, from the very small to the very large, and excel at sanitary applications due to their ability to incorporate sanitary-friendly liners. While ultrasonic and DP flowmeters are alternatives for some applications, magnetic flowmeters have a large installed base and there does not seem to be a strong movement to alternative technologies.

Given the strength of magnetic flowmeters in the above industries, a frontier of research is to focus on developing more products specifically designed for these industries. One option is to incorporate a control valve with a magnetic flowmeter to provide control as well as measurement. Another is to look more closely at the needs of the beverage industry, including breweries and microbreweries, which already use both magnetic and Coriolis meters. Magnetic flowmeters do compete with Coriolis meters in the small line sizes, but magnetic flowmeters have a price advantage there. Developing customized solutions for applications in the five industries mentioned above that may incorporate additional instrumentation could be a winning solution for a magnetic flowmeter company that wants to gain market share.

6 Ultrasonic Flowmeters

Ultrasonic flowmeters have been gaining acceptance over the last decade as end-users come to understand and appreciate the technology – although some are just now discovering the advantages and potential of ultrasonic flow measurement.

The ultrasonic flowmeter market is still a relatively new technology. Tokyo Keiki first introduced ultrasonic flowmeters to commercial markets in Japan in 1963. In 1972, Controlotron introduced the first clamp-on ultrasonic flowmeter to the United States. In the late 1970s and early 1980s, Doppler flowmeters began to be used.

Because ultrasonic flowmeters were not well understood at first, they were often misapplied. As a result, many users got a bad impression of the meters during this time. It was not until the 1990s that ultrasonic flowmeters began to be widely used for industrial applications.

The rapidly expanding market for gas flow measurement is one of the major reasons for strong projected growth in the ultrasonic flowmeter market. Energy, including energy conservation, and other markets have the potential to create even more demand, particularly as the technology improves to enable new applications.

ADVANTAGES

Ultrasonic flowmeters feature high accuracy, high reliability, high turndown ratios, long service life, low maintenance, relatively low cost, valuable diagnostics, no moving parts, and redundancy capabilities. Clamp-on ultrasonic flowmeters, in particular, can offer redundancy by providing an easy check of an inline meter. In addition to the traditional advantages, suppliers are significantly improving accuracy, sensitivity, and reliability.

Ultrasonic flowmeters have a distinct advantage over other flowmeters:

Unlike Coriolis meters, ultrasonic flowmeters do very well in large pipe sizes.

In large-size, natural gas pipeline applications, ultrasonic flowmeters have the advantage over turbine and differential pressure (DP) flowmeters of being highly accurate, non-intrusive, and highly reliable over time, with no moving parts to wear. They also have an advantage over DP flowmeters in that they are largely non-intrusive, with the exception of insertion types.

Ultrasonic flowmeters have an advantage over magnetic flowmeters in that they can be used to measure the flow of nonconductive liquids, gases, and steam.

Ultrasonic flowmeters have an advantage over vortex flowmeters in that they can measure low flows better than vortex meters.

Further sweetening the pot is the fact that average ultrasonic prices are holding their own or even declining. In comparison, the average price for Coriolis flowmeters has had upward pressure due to the recent introductions of large-line size models in the 12"–16" diameter range (Table 6.1).

DOI: 10.1201/9781003130017-6

TABLE 6.1
Advantages and Disadvantages of Ultrasonic Flowmeters

Advantages	Disadvantages
High accuracy	High initial cost
Approved for custody transfer	Dirt and fluids impact performance (increased bearing friction) and measurement accuracy (distorted dimensions)
High turndown/rangeability – measure low and high pressures	Noise even beyond human hearing range interferes with detection of sonic pulses
Repeatability	Pipe walls can interfere with ultrasonic signals for clamp-on meters
Can clamp on to a pipe with no penetration	Build-up on the inside pipe wall can reduce internal pipe diameter and affect measurement accuracy
Tolerate extreme temperatures	Insertion ultrasonic meters have limited accuracy
Self-diagnosing; once calibrated, assessments can determine measurement shifts, requiring less frequent calibration	
Long-term reliability	
Low maintenance (no moving parts to replace or lubricate)	

A RANGE OF APPLICATIONS

Ultrasonic flowmeters are used in upstream applications for allocation metering, for measuring gas and oil from test and production separators, for check metering, and for other applications. A major use of ultrasonic flowmeters is in the midstream segment for custody transfer of natural gas.

Probably the single most important factor in the recent growth of ultrasonic flowmeters in the past 20 years has been the rapid growth in the market for multipath ultrasonic meters for custody transfer of natural gas. Multipath ultrasonic meters have three or more paths. The benefit of having multiple paths is that flow is measured at more points in the flowstream. This enhances the accuracy of the measurement. In 1998, the American Gas Association (AGA) approved the use of ultrasonic flowmeters for custody transfer applications. Since that time, suppliers have researched multipath meters and brought out new products.

Ultrasonic flowmeters are also being more widely used to measure process gas and flare gas. Insertion meters are used to measure flare gas in stacks, and ultrasonic flowmeters are used more widely in the chemical and refining industries.

In other areas, the increased use of battery power supplies opens up additional opportunities for ultrasonic flowmeters in the water and irrigation industries as well as other applications. Battery-powered flowmeters offer low-energy consumption independent of the main power grid and can help reduce prices on the worldwide market.

Another important application for ultrasonic flowmeters is check metering – verifying another flowmeter's readings. Some clamp-on meters are used for this purpose since they can be conveniently moved from one meter to another. (Clamp-on technology is unique to ultrasonic meters.) Inline ultrasonic flowmeters are also used for check metering.

ULTRASONIC FLOWMETER COMPANIES

Baker Hughes Company

Baker Hughes, an energy technology company with operations in more than 120 countries, provides oilfield products and services for the oil and gas industry. The company is primarily involved in drilling, surface logging, pipeline operations, petroleum engineering, and fertilizer solutions. It also supplies gas turbines, valves, actuators, pumps, generators, motors, and flowmeters.

History and Organization

Baker Hughes' flowmeter business dates back to Edmund Carnevale and David Chleck, who started Panametrics in 1960, and introduced ultrasonic flowmeters to measure gas flow in the 1970s. However, the market at the time was not yet ready to accept this technology, so in the 1980s Panametrics turned its attention to measuring flare gas. Eventually, Panametrics brought out the GN868 Series for measuring the flow of natural gas. In July 2002, General Electric's Power System's division purchased Panametrics for US$250 million and added the firm and its approximately 1,000 employees to GE's Energy Services Division.

In 2004, the Panametrics unit was one of seven groups put together to form GE Infrastructure Sensing. In 2005, this group became GE Sensing. In 2006, GE Sensing and GE Inspection Technologies merged to become GE Sensing & Inspection Technologies.

In 2009, GE Infrastructure, Sensing & Inspection Technologies and GE Optimization & Control merged to become GE Measurement & Control Solutions. In addition to the Panametrics line, the company created its instrumentation product lines through acquiring Dresser, Druck, General Eastern, Kaye, NovaSensor, Pressurements, Protimeter, Rheonik, Ruska, SI, Thermometrics, Telaire, and others.

In 2015, GE Measurement & Control became part of GE Oil & Gas. In the following year, GE Oil & Gas announced a merger with Baker Hughes, creating Baker Hughes, a GE Company (BHGE). This new company included the former GE Measurement & Control. In 2018, however, GE, with a new business focus in mind, sold off a percentage of BHGE and expressed its intention to further reduce its share. By this time BHGE was already an essentially independent operation.

In early 2019, BHGE announced plans to market its flowmeter product line, along with the original product lines from Panametrics, Inc., under the PANAMETRICS name. PANAMETRICS was to remain a part of the future of BHGE and not be a separate business unit. On October 19, 2019, BHGE announced that it had a new name, Baker Hughes Company and would be known as Baker Hughes. It also immediately began trading on the New York Stock Exchange under the symbol "BKR." Today, the Panametrics product line is branded as Panametrics, a Baker Hughes business.

Ultrasonic Flowmeters

Panametrics groups its transit time ultrasonic flowmeters into four categories: precision flare gas flowmeters; process flowmeters for liquids, gases, and steam; custody transfer flowmeters; and portable liquid and gas clamp-on flowmeters. In addition to clamp-on meters, Panametrics offers inline and insertion ultrasonic meters.

The latest entrant into the Panametrics family of high accuracy meters, Sentinel LCT8, has eight paths, some of which are used to cancel the effects of swirl and other disturbances. The Sentinel LNG uses Bundle Waveguide™ technology to protect the ultrasonic transducers from cryogenic temperatures. Other Sentinel series ultrasonic flowmeters are used for liquid custody transfer, including hydrocarbon liquids.

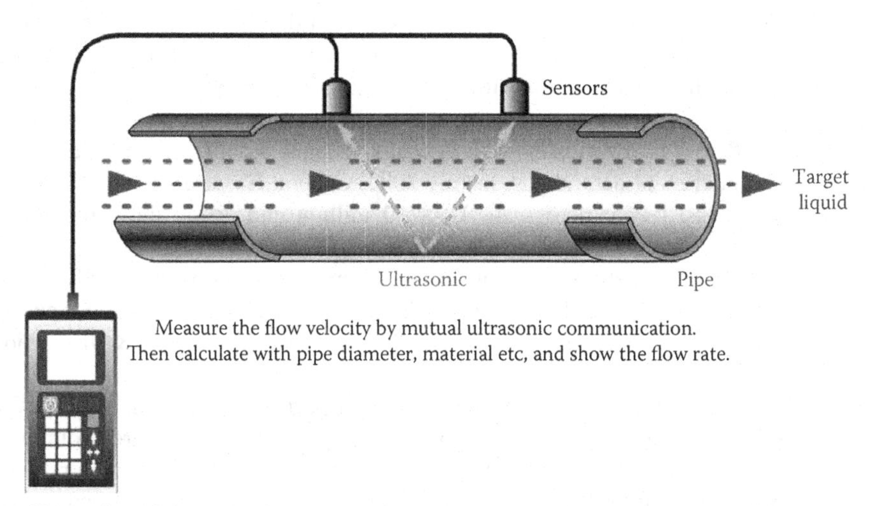

An illustration of the path of an ultrasonic signal. (Source: Courtesy Tokyo Keiki.)

SICK AG

SICK AG is a technological and market leading producer of industrial sensors and sensor solutions to three primary business automation fields: factory, logistics, and process. An important aspect of SICK's success in process automation has been the firm's emergence as an important player in the ultrasonic flowmeter market and one of the top suppliers for custody transfer. SICK also offers a wide range of sensor, analyzer, and switch products (e.g., photoelectric, safety light curtains, gas analyzers, and bar code scanners).

History and Organization

The company's modest beginnings were near Munich, where the self-employed Dr. Erwin Sick worked on a variety of optical-electronic devices. Sick's first real technical breakthrough was the development of a commercially successful accident prevention light curtain, an industrial safety device, in 1952. After Erwin Sick's death in 1988 at the age of 79, his widow, Gisela Sick, began running the company and later became an honorary member of the Supervisory Board.

In 1996, the company converted from a GmbH, a kind of private limited company, to an Aktiengesellschaft, a joint stock company. In 1999, it issued its first employee shares domestically and abroad. In 2004, SICK Engineering became a subsidiary of the newly founded SICK MAIHAK GmbH. By 2005, the company had more than 4,000 employees and almost 50 subsidiaries in more than 20 countries, and numerous affiliated companies and agencies.

As part of SICK's continued growth and development, the Flow Business Unit, formerly SICK Engineering GmbH, was elevated to a division of the SICK Group in January 2011.

The business center for Flow Measurement began in 1991 as SICK Engineering GmbH, a SICK subsidiary near Dresden. In 1994, the subsidiary introduced FLOWSIC101, its first flow monitor, followed by the FLOWSIC600 gas meter in 2003. In 2004, SICK Engineering became a subsidiary of the newly founded SICK MAIHAK GmbH. In January 2013, SICK MAIHAK transferred all its operations to SICK AG and retired the Maihak brand.

Ultrasonic Flowmeters

SICK provides custody and non-custody ultrasonic gas flowmeters that target the oil and gas as well as the chemical and refinery industries. SICK provides solutions for a wide variety of measuring tasks, including calculating volume flows in processes, environmental and emission monitoring of plants, and test stands in the automotive industry. This product portfolio is focused on the entire value chain of natural gas flow, from production facilities to municipal utilities and industrial consumers. SICK also delivers metering skids and metering runs as customized turnkey solutions that incorporate ultrasonic technology for custody and non-custody gas transfer.

The entire FLOWSIC product family is characterized by a unique flow-optimized sensor design that reduces flow noise and signal drift to a minimum even at very high gas velocities. Enhanced signal processing and highly efficient transducers result in high time resolution of the signal and, therefore, accurate measurement even at very low gas flowrates.

FLOWSIC600-XT, SICK's newest generation of ultrasonic gas and steam flowmeters, is used in both low- and high-pressure pipelines, process and control tasks, and custody transfer applications.

HOW THEY WORK

There are two main types of ultrasonic flowmeters, transit time and Doppler:

Transit Time Ultrasonic Flowmeters

A *transit time ultrasonic flowmeter* has both a sender and a receiver. It sends two ultrasonic signals across a pipe at an angle: one with the flow, and one against the flow. The meter then measures the "transit time" of each signal. When the ultrasonic signal travels with the flow, it travels faster than when it travels against the flow. The difference between the two transit times is proportional to flowrate.

Transit time ultrasonic flowmeters are distinguished according to the number of "paths" they have. A path is simply the path or track of the ultrasonic pulse as it travels across the pipe and back again. Many ultrasonic flowmeters are single or dual path, meaning that they send either one or two signals across a pipe and back. Typically, there are two transducers for each path; one is a sender and one is a receiver (Figures 6.1 and 6.2).

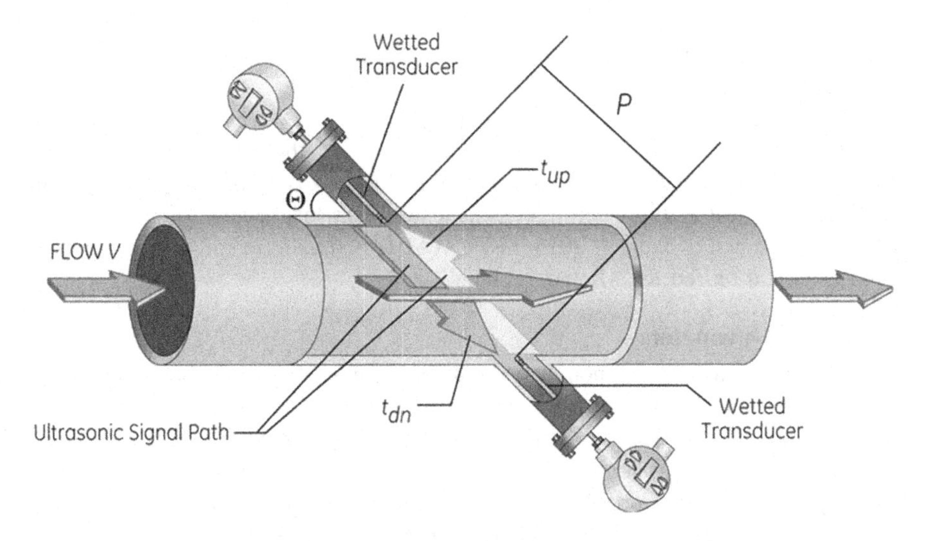

FIGURE 6.1 The operating principle of a transit-time based ultrasonic flowmeter. (Source: Panametrics, a Baker Hughes business.)

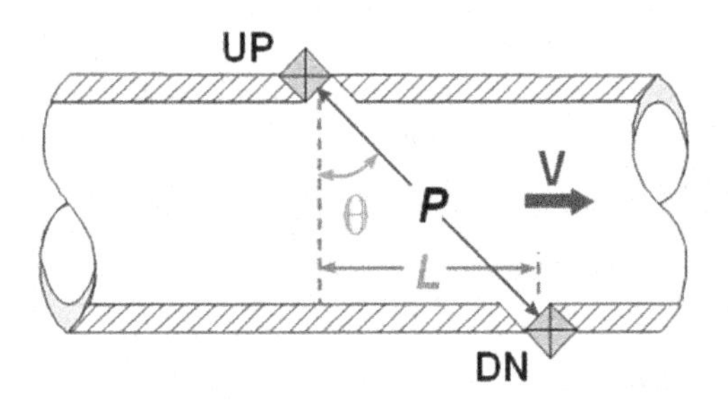

FIGURE 6.2 Upstream and downstream transit path for wetted applications. (Source: Panametrics, a Baker Hughes business.)

If we know the following parameters:

- The path length P
- Sound speed c
- Flow velocity V
- Beam angle q

The travel time for the upstream and downstream signals, t_{up} and t_{dn}, can be calculated as:

$$t_{up} = \frac{P}{c - V \sin \theta}$$

$$t_{dn} = \frac{P}{c + V \sin \theta}$$

The transit time method measures t_{up}, t_{dn}, and Δt to obtain the velocity of the flow:

$$V = k\frac{P^2}{2L}\left(\frac{t_{up} - t_{dn}}{t_{up}t_{dn}}\right)$$

with k = Reynolds correction factor.

Doppler Flowmeters

Doppler flowmeters also send an ultrasonic signal across a pipe. Instead of tracking the time the signal takes to cross to the other side, a Doppler flowmeter relies on having the signal deflected by particles in the flowstream. These particles are traveling at the same speed as the flow. As the signal passes through the stream, its frequency shifts in proportion to the mean velocity of the fluid. A receiver detects the reflected signal and measures its frequency. The meter calculates flow by comparing the generated and detected frequencies. Doppler ultrasonic flowmeters are used with dirty liquids or slurries. They are not used to measure gas flow.

The Difference between Paths and Chords

End-users perceive, rightly or wrongly, that the more paths or chords a flowmeter has, the more accurate it is, and the more diagnostic information it is capable of gathering. Consequently, some suppliers have come to emphasize the number of chords their meter has over the number of paths, since in many cases this is a higher number.

In reality, the difference between chords and paths is far from obvious and requires some real studying to understand.

Path versus Chord

In flowmeter terminology a path is defined as the route of travel between two ultrasonic transducers. The term "path" is critical in ultrasonic technology because many ultrasonic flowmeters have been developed with multiple paths. Some ultrasonic meters have a single path, requiring one pair of transducers, and some have dual paths, requiring two transducer pairs. An important group of ultrasonic flowmeters have three or more paths and are called multipath. Many of these multipath meters are used for custody transfer applications.

Another term that is now in common use is "chord." Mathematically speaking, a chord is a straight line within a circle whose points lie on the circumference. However, the term "chord" is also used by some ultrasonic manufacturers to refer to the route of travel between two transducers. In this way, a chord is like a path. However, a chord is considered to be the route of travel between a transducer and a wall or reflector when the signal is bounced off a wall or a reflector. So, in this sense, an ultrasonic signal that bounces off a wall or reflector to a receiving transducer has one path and two chords. One chord is the path of the signal from transducer A to the pipe wall or reflector, and the second chord is the path of the signal from the pipe wall or reflector to transducer B.

This can be made clearer by distinguishing between a direct path and a reflected path. In a direct path, the signal goes from one transducer to another without being reflected. In a reflected or bounced path, the signal from one transducer is reflected or bounced off a reflector or the pipe wall on the way to the second transducer. Unlike a reflected path, a direct path is considered to have only one chord.

Is Chord "Marketing Speak"

Some people believe that the term chord is "marketing speak" and that the terms "chord" and "path" mean the same thing. Whether or not the term *chord* is "marketing-speak," it is clear that *path* and *chord* do not mean the same thing. A "chord" is an ultrasonic signal that travels from one side of a pipe to the other. A "path" is an ultrasonic signal that typically travels from one side of the pipe to the other and back. So, in many cases an ultrasonic flowmeter has more chords than paths. For example, Elster's Q.Sonic Plus ultrasonic meter has six paths and 16 measurement chords.

SUPPLIERS' INTERPRETATION OF PATHS AND CHORDS – AND WHAT THIS MEANS FOR ULTRASONIC FLOWMETERS

In an effort to understand what all this means, we asked a number of suppliers to give their interpretation of paths and chords, and what this difference means for ultrasonic flowmeters. Here are their responses.

KROHNE

KROHNE speaks of beams and chordal paths, which in "KROHNE-speak" seems to refer to the "slice" of the flowstream they represent. A single chordal path (one path = two chords, one with flow, one against flow) is also spoken of as a single beam. This generally slices through the center of the pipe at its widest point. KROHNE adds more chordal paths or beams that do not cut through the center of the pipe in order to better understand the flow profiles of the fluid. This also helps to maintain accuracy throughout a broader operating range than possible in a single path/beam device. While it might not be directly represented in the accuracy specifications, more paths (more slices) result in better and consistent performance in a wider range of fluids or flowing conditions (Figure 6.3).

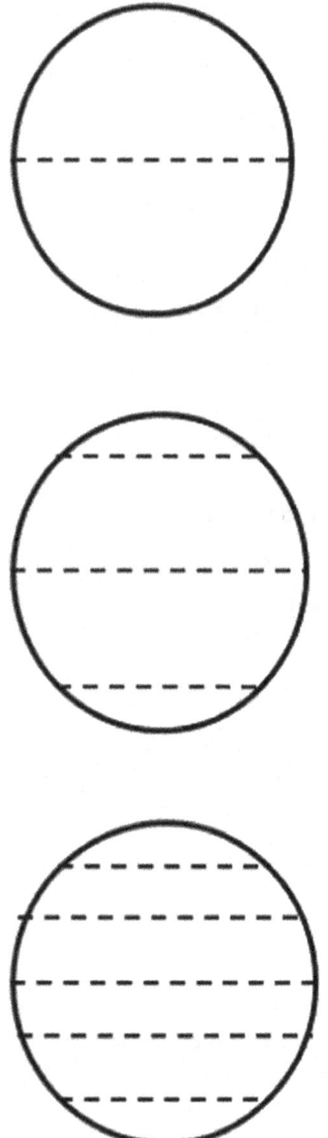

FIGURE 6.3 KROHNE single beam, three beam, and five beam path configurations. (Source: Courtesy KROHNE.)

Cameron

From Cameron, Dr. Gregor Brown supplied the following definitions:

Transducer: Device for sending and receiving ultrasonic signals.

Path: Route of travel through the fluid between two transducers (note: both transducers on all paths ALWAYS play the role of both transmitter and receiver in turn, i.e., a signal is sent along the path from A to B and then the roles are reversed, and a signal is sent from B to A).

Diameter: Line joining two points on the circumference of a circle, passing through the center point.

Diametric path: Path that lines up on a diameter in a cross-sectional view of the meter.

Chord: Line joining two points on the circumference of a circle.

Chordal path: Path that lines up with a chord in a cross-sectional view of the meter.

Direct path: Path that connects two transducers without using any reflection points. Direct paths can be diametric or chordal.

Reflected or bounce path: Path that connects two transducers via one or more reflection points. Reflected paths can be diametric or chordal (note: depending on where you wish the reflected path to go, the reflection point can either be the pipe wall itself, or a reflector can be used to send the path off at a different angle).

Traverse: Segment of a path connecting two points at the pipe wall – the points can either be defined by a transducer or a reflection point, and the traverse may be diametric or chordal.

Single bounce path: Path with one reflection point and two traverses.

Double bounce path: Path with two reflection points and three traverses.

Triple bounce path: Path with three reflection points and four traverses.

Emerson

Emerson supplied several definitions from ISO 17089:

3.1.3.2 Acoustic path: Path traveled by an acoustic wave between a pair of transducers.

4.3.3.3 Commonly used multipath cross-sectional configurations (see Figure 6.4).

Elster

A direct path is equal to a chord (or "chordal path"). A single reflection path consists of two chords and a double reflection path consists of three chords.

Taking a mid-radius double reflection path of our Q.Sonic-plus as an example, it is three times longer compared to a direct mid-radius path (assuming the same angle).

The fact that the travel time of the ultrasonic signal is longer and therefore the time difference measured, with and against the flow direction, is also longer, results

4.3.3.3 Commonly used multipath cross-sectional configurations

The cross-sectional configuration is important as this dictates what information about the axial velocity distribution is available for the computation of the average axial velocity. Selections of commonly encountered cross-sectional configurations are shown in Figure 7.

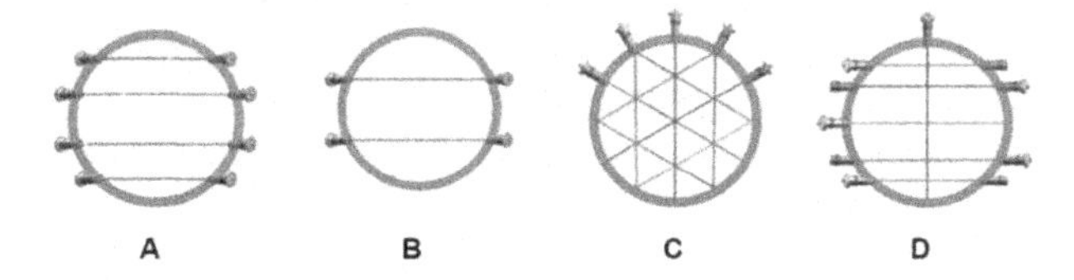

FIGURE 6.4 Some typical cross-sectional acoustic path configurations. (Source: Courtesy Emerson Automation Solutions.)

in a higher resolution and typically a better flow measurement accuracy and repeatability especially for the lower flows (Figure 6.5).

Also, the ultrasonic signal "collects" more gas velocity information when traveling longer and at different positions through the gas, which again may result in a more accurate flow measurement.

To talk about chords instead of paths is simply to visualize this situation. In Elster's case, a six-path Q.Sonic-plus has 16 chords.

Accusonic

Accusonic views the proper definition of a path in a fairly restrictive manner. Specifically, each path must incorporate its own independent "delta t" measurement. As such, an 18-path transit-time flowmeter comprises 18 paths with each path independently measuring "delta t" on a continuous (repetitive) basis. Time of transit in the forward and reverse directions is measured along each path, and "delta t" is computed as the difference in these two measured times along each path. A path may be comprised of one or more chords and both paths and chords may be parallel or non-parallel.

The illustrations on the next pages are examples of two different types of 18-path transit-time flowmeters, one with two planes of nine parallel paths (symmetrically crossed) and one with 18 non-parallel paths (the end-view image only shows half the total number of paths).

What It Means

Suppliers appear to agree on the definition of "chord" and "path." A path is the area taken up by an ultrasonic signal that passes back and forth between two transducers. If the path is direct, then there is only one path and one chord between the two transducers. If the path is bounced or reflected, there is one path but there are two or more chords between the two transducers (Figure 6.6).

While suppliers agree on definitions, they seem to have different views about the value of bounced or reflected paths. Some say that added length in the chords in

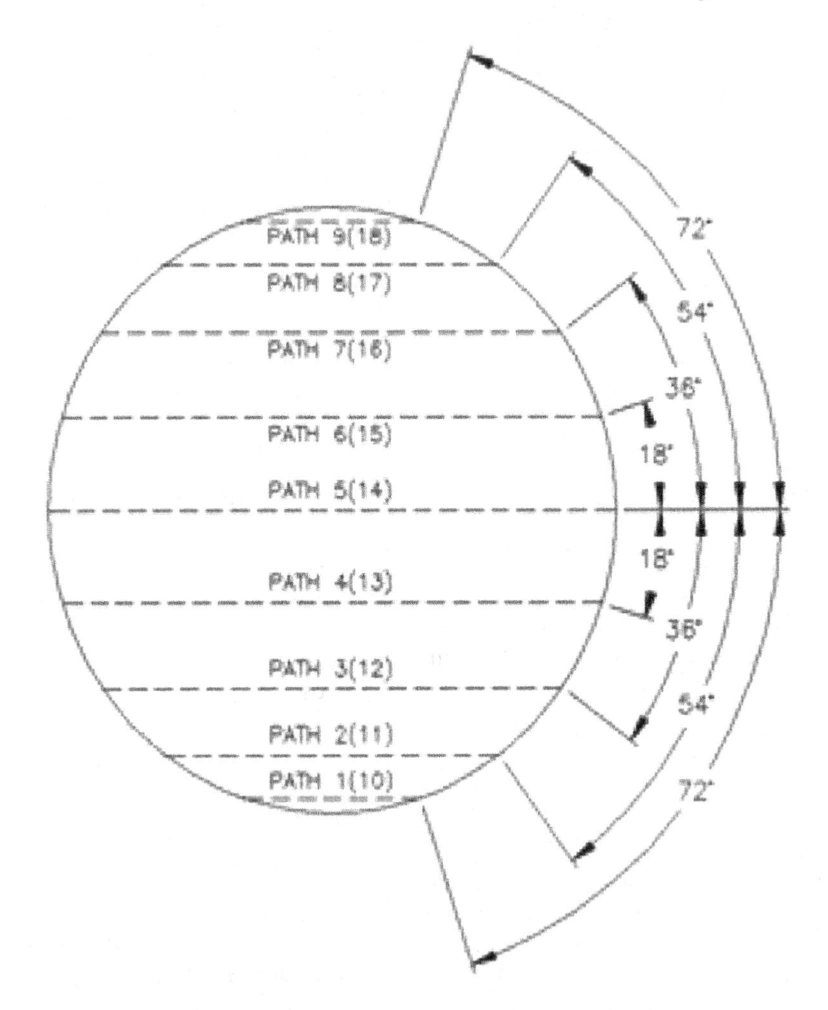

FIGURE 6.5 18-path flowmeter with two planes of nine parallel paths.

reflected or bounced paths provides additional diagnostic information and enhanced performance. It can also provide greater measurement accuracy and repeatability, according to some suppliers.

Other suppliers suggest that talk of chords by suppliers who emphasize the number of chords over paths are engaged in "marketing-speak" and that end-users may interpret the number of chords to be the same as the number of paths. These suppliers prefer to talk about paths rather than chords.

Part of the problem here is that the distinction between paths and chords is not so easy to understand. Supplier literature typically does not address this issue, so end-users have to rely on journal articles or conferences to grasp the difference. And many probably do not have the time or interest to follow this discussion in depth.

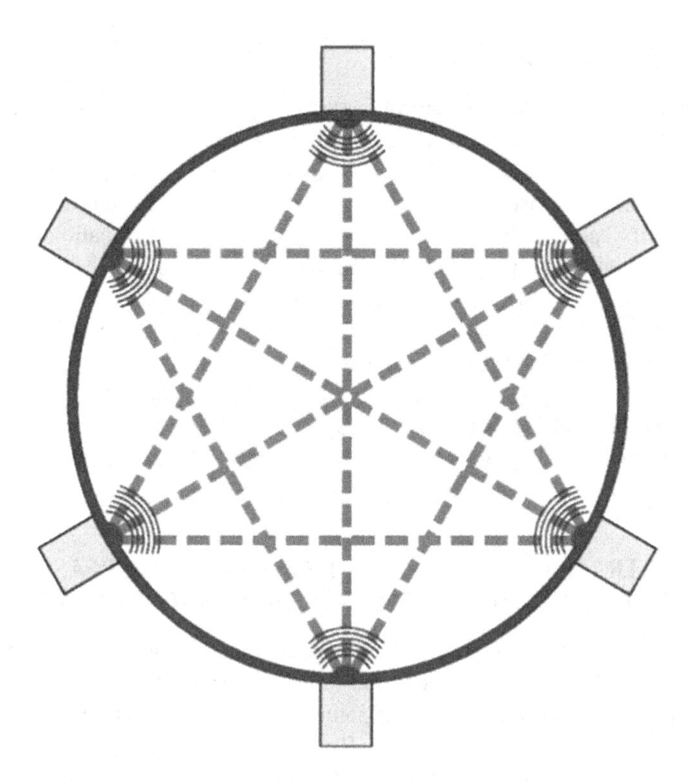

FIGURE 6.6 Flowmeter with 18 non-parallel paths (end-view image only shows half the total number of paths). (Source: Courtesy Accusonic.)

One Solution: Talk about Chords and Paths Together

Perhaps an independent testing agency could resolve the question whether having additional length in chords provides genuine diagnostic information. If having multiple chords in bounced paths really provides an edge in performance, accuracy, or flow profile analysis, then there is good reason to talk about the number of chords in an ultrasonic flowmeter. It is not "marketing-speak" to talk about features that genuinely enhance performance, any more than it is "marketing-speak" to talk about reducers in vortex meters or sanitary Coriolis meters.

On the other hand, suppliers who talk only about chords and don't mention paths might appear to be taking advantage of people's lack of knowledge about the terms. *So, one solution is to talk about the number of chords, but also talk about the number of paths at the same time*. As an example, Elster talks about the Q.Sonic-plus as having six paths and 16 chords. This makes it clear how many chords are in the meter, and that there are actually fewer paths than chords.

Everyone Has a Responsibility Here

It is certainly part of the responsibility of suppliers not to use words in a misleading way, and to try to be accurate in describing their products. At the same time, it is up to end-users to learn the terminology that describes the products they use, if they want to get maximum use out of them. It is probably wise for end-users to have a "trust but verify" approach to manufacturer claims. But the verification part requires knowledge, which in turn requires education.

End-users should also look to the results of independent tests on different flowmeters to judge which ones have the best performance. Often the specifications in a brochure reflect what happens in laboratory conditions, not what happens in the real world. All this is part of evaluating what is the best flowmeter solution for a given application. And the better end-users understand the terminology employed to describe the flowmeters they are evaluating, the better the chance that they will be able to choose wisely when selecting a flowmeter.

CUSTODY TRANSFER OF NATURAL GAS IS AN INCREASINGLY IMPORTANT MEASUREMENT FOR ULTRASONIC FLOWMETERS

Probably the single most important factor in the recent growth of ultrasonic flowmeters in the past 20 years has been the rapid growth in the market for multipath ultrasonic meters for custody transfer of natural gas. The initial surge in growth began in 1995, when Groupe Européen de Recherches Gazières (GERG), published Technical Monograph 8 – *Present status and future research on multipath ultrasonic gas flowmeters*. This technical document laid out criteria for using these instruments to measure natural gas flow for custody transfer. Its publication gave a major boost to the sales of multipath ultrasonic flowmeters for natural gas in Europe.

After the publication of the GERG document in Europe, ultrasonic suppliers worked with the American Gas Association (AGA) to obtain approval of a similar document in the United States. This resulted in the publication of AGA-9 in June 1998. AGA-9 lays out criteria for buyers and sellers of natural gas to follow when using ultrasonic flowmeters for custody transfer of natural gas. The publication of AGA-9 resulted in another major increase in the sales of multipath ultrasonic flowmeters for natural gas. This increase is continuing through today, even though the demand for custody transfer applications was down somewhat due to reduced oil and gas exploration and production as a result of the COVID-19 pandemic in 2020 and 2021.

Ultrasonic flowmeters are also being more widely used to measure process gas and flare gas. The use of ultrasonic meters for gas flow measurement has evolved substantially from the late 1970s and early 1980s, when this was first attempted. The use of wetted sensors provides greater accuracy. Insertion meters are used to measure flare gas in stacks, and ultrasonic flowmeters are used more widely in the chemical and refining industries. While the growth of ultrasonic meters to measure process and flare gases is not as rapid as is the growth of multipath meters for custody transfer of natural gas, it is still an important factor in the overall growth of ultrasonic meters.

The ultrasonic flowmeter market is still a relatively new technology. Tokyo Keiki first introduced ultrasonic flowmeters to commercial markets in Japan in 1963. These were clamp-on flowmeters. In 1972, Controlotron introduced the first clamp-on ultrasonic flowmeters to the United States. In the late 1970s and early 1980s, Doppler flowmeters began to be used. Because they were not well understood, they were often misapplied. As a result, many users got a bad impression of ultrasonic flowmeters during this time. It was not until the 1990s that ultrasonic flowmeters began to be widely used for industrial applications.

The ultrasonic flowmeter market is highly fragmented with more than 60 suppliers worldwide. For comparison purposes, there are less than half that number of Coriolis suppliers worldwide, and more than 50 suppliers of magnetic flowmeters. Many of the ultrasonic flowmeter companies are suppliers of clamp-on and/or insertion meters alone. A number of these ultrasonic flowmeter companies can also be identified as niche suppliers who sell mainly to a specific geographic region, or to one or more particular industrial sectors. And part of the reason for the growth in ultrasonic meters is the sheer number of suppliers (Figure 6.7).

Growth Factors for the Ultrasonic Flowmeter Market

- Ultrasonic flowmeters have gained the trust of end-users.
- Ultrasonic flowmeter technology compares well to other flowmeter types.
- Signal processing technology is improving.
- The intrinsic advantages of ultrasonic technology.
- Multipath ultrasonic flowmeters are used for custody transfer.
- Transit time flowmeters have expanded their capability to measure fluids with particles.

FIGURE 6.7 Transducers on an ultrasonic flowmeter.

- Flow calibration facilities are now more available.
- Flow measurement redundancy is now more important.
- Clamp-on models hit their stride in very large pipe diameters.
- The worldwide business of regulating greenhouse gas emissions.

Ultrasonic Flowmeters Have Gained the Trust of End-Users

When ultrasonic flowmeters were introduced in the 1970s, the technology was not well understood. As a result, ultrasonic flowmeters were frequently misapplied, leading to bad results. Ultrasonic flowmeters were initially subject to some installation challenges that are now mostly avoidable, but these led to ultrasonic flowmeters gaining an unfavorable reputation and creating doubts about the technology's promised performance among some users. One example of this doubt were issues related to electrical noise produced by other process devices (e.g., pressure drops at control valves) and its propagation upstream and downstream from the source. If left untreated, this noise could interfere with the flowmeter's signal detection system and create intermittent inaccuracies in transit time measurement. In addition, some users experienced unsatisfactory results with Doppler meters, which have never been as accurate as transit time meters.

Over the years, however, both Doppler and transit time technologies have substantially improved. In fact, successful user experiences with ultrasonic flowmeters have served as paradigm cases for the rest of the industry. Many more suppliers have entered the field, and these suppliers have helped educate users about the use and advantages of ultrasonic flowmeters. This better understanding makes companies more likely to install and use ultrasonic flowmeters for optimal benefit.

Ultrasonic Flowmeter Technology Compares Well to Other Flowmeter Types

Inline ultrasonic flowmeters have a host of benefits that have made the technology a popular choice for a wide range of flow measurement applications in gas, liquid, and steam environments. These advantages include high flow measurement accuracy, high reliability, high turndown ratios, competitive pricing, no moving parts, low maintenance, valuable diagnostics, bi-directional flow measurement, redundancy capabilities, and long service life. Their accuracy claims can also be verified by accredited calibration standards according to ISO 17025.

Ultrasonic flowmeters have distinct advantages over each of the other new-technology meters. Unlike Coriolis meters, ultrasonic flowmeters do very well in large pipe sizes. Over half of the Coriolis meters sold today are for pipe sizes two inches or less. While some Coriolis meters have successfully been used in four-inch and six-inch lines, they begin to become unwieldy and quite more expensive in sizes over two inches. Size can actually be an advantage for inline ultrasonic flowmeters, since larger pipes provide more distance for the ultrasonic signal to cross. For pipes six inches and larger, ultrasonic flowmeters are in most cases a better choice than Coriolis meters.

One important change in the Coriolis market is that Coriolis suppliers are manufacturing Coriolis flowmeters for larger line sizes. Today, Endress+Hauser, Micro Motion, KROHNE, Rheonik, and Shanghai Yinuo are among a group of manufacturers making Coriolis flowmeters for line size diameters ranging from 8 to 16 inches. These meters mainly target the fast-growing oil and gas industries, and many are designed for custody transfer applications. While the price of these meters can easily reach into the $75,000 range or more, they are highly accurate and can offer significant return on investment when dealing with high volumes of high-value fluids. Because of their operating principle, Coriolis flowmeters can measure liquid flows more easily than gas flows, but some of these large size meters can also be used to measure the flow of natural gas.

Despite the advent of these larger line size Coriolis flowmeters, inline ultrasonic flowmeters still have an edge over Coriolis technology. Many of the ultrasonic flowmeters used for custody transfer of natural gas range in size from 12 to 42 inches and up. Coriolis flowmeter suppliers have yet to engineer a practical design for Coriolis flowmeters in these line sizes.

The main competitors to ultrasonic meters for natural gas pipeline applications in the larger line size range are turbine and differential pressure (DP) flowmeters. Here, ultrasonic flowmeters have the advantage of being highly accurate, non-intrusive, and highly reliable over time, with no moving parts to wear. Ultrasonic flowmeters can also detect air/gas mixtures and raise an alarm; mechanical flowmeters register gas/air mixtures as liquid and consequently introduce an error into the measurement. Because they have been in use for many more years than ultrasonic meters, turbine and DP flowmeters presently maintain an advantage in the size of their installed base and user familiarity. But many companies are now opting for ultrasonic meters over turbine and DP meters, especially for new applications.

For a long time, DP flowmeters also had the advantage of approvals from the American Petroleum Institute (API) and the AGA that helped them establish a large installed base in the oil and gas industry. This advantage has diminished as ultrasonic flowmeters have gained their own sets of approvals from international bodies.

Inline ultrasonic flowmeters have an advantage over DP flowmeters in that they do not interfere with the process flowstream. The intrusiveness of DP flowmeters varies with the primary element used. Orifice plates, for example, are very intrusive as they are designed to cause pressure drop by hindering the flow. Orifice plates can also get knocked out of position or otherwise damaged by the fluid flow, which requires DP flowmeters with orifice plates to be recalibrated frequently.

Accurate measurement of the transit time is critical if an ultrasonic meter needs to meet performance requirements established in AGA-9. (Single-path ultrasonic meters have settled into gas distribution applications where the volumes and pressures are considerably less than for their multipath counterparts.)

AGA-9 was a significant development in the use of ultrasonic flowmeters. It followed Technical Monograph 8, a 1995 report by the Groupe Européen de Recherches Gazières (GERG) – a technical document that laid out the criteria for using ultrasonic flowmeters for custody transfer of natural gas. First released in June 1998 (the most recent edition was released in 2022), Report Number 9 further described the criteria for using ultrasonic flowmeters for custody transfer of natural gas. Report Number 9 introduced ultrasonic flowmeters to the important custody transfer application space and continues to give a major boost to the ultrasonic market in the oil and gas production and transportation industry.

Ultrasonic flowmeters have an advantage over magnetic flowmeters in that they can be used to measure the flow of nonconductive liquids, gases, and steam. Magnetic flowmeters have very limited use in oil and gas production, transportation, and refining sectors because petroleum-based liquids are nonconductive. For the most part, magnetic flowmeters simply cannot be used to meter hydraulic fluids, oil or natural gas, process gas, flare gas, or steam – and this is one of the most important reasons why the ultrasonic flowmeter market is growing so much faster than the magnetic flowmeter market. Magnetic flowmeters are simply unable to participate in the fast-growing gas and steam flow measurement markets.

Ultrasonic flowmeters have an advantage over vortex flowmeters in that they can measure low and zero flows. Vortex meters have a difficult time with low flows because if the flow is too low, it may not generate sufficient vortices for measurement purposes. Vortex flowmeters also have a difficult time registering zero flow. Inline ultrasonic flowmeters are also non-intrusive, and do not introduce pressure drop or affect process throughput. While vortex meters are not as intrusive as orifice plate meters, their bluff bodies can get knocked out of position if there are sufficient impurities in the flowstream.

The rapidly expanding market for gas flow measurement using ultrasonic flowmeters is one of the major reasons for the strong projected growth in the ultrasonic flowmeter market. This market includes natural gas, process gases, and flare gas. Of these three, the market for natural gas flow is expected to show the strongest growth. This market involves the use of multipath ultrasonic flowmeters for custody transfer. However, ultrasonic flowmeters are now being more widely used in the chemical and refining industries to measure process gases. These meters are an alternative to magnetic flowmeters that cannot meter nonconductive liquids or gas. Insertion and inline ultrasonic meters will continue to be used to meter flare gas in large pipes.

Signal Processing Technology Is Improving

It was mentioned earlier that the first ultrasonic flowmeter models were subject to measurement inaccuracies due to noise produced by other devices associated with the process flow that would interfere with the ultrasonic signals being sent and received by the transducers. In addition, if a clamp-on ultrasonic flowmeter was not properly installed and maintained, attenuation of the ultrasonic signal could occur at the interfaces between the clamp-on transducers and the outside pipe walls, and between the inside pipe walls and the fluid. Simple fixes such as being able to reliably maintain clamp-on transducer connections to the pipe wall meant that the ultrasonic waves would now always reach the fluid to be measured.

Many of these physical connection issues have been addressed over the years through more consistent installation routines and minimizing or eliminating any maintenance requirements. What has also been critically important are improvements to the ultrasonic signal technology itself. Improvements in transducer material and design, increases in available dB levels of transmission, and reductions in signal attenuation through better electronics have all contributed to better measurement accuracy performance over time.

Multipath Ultrasonic Flowmeters Used for Custody Transfer

One important technological improvement has been the development of multipath transit time flowmeters, which use more than one ultrasonic signal or "path" in calculating flowrate. Multipath ultrasonic flowmeters use multiple pairs of sending and receiving transducers to determine flowrate. The transducers alternate in their function as sender and receiver over the same path length. The flowrate is determined by averaging the values given by the different paths, providing greater accuracy than single path flowmeters. By strategically locating the positioning of these paths, the flowmeter measures flow at multiple points of cross section within the flowstream's profile, leading to greater accuracy.

Today, close to half of all ultrasonic flowmeters sold for industrial process control are purchased to measure gas flow. Of these, roughly 35 percent are used in natural gas custody transfer applications, and only multipath flowmeters are currently approved for this purpose. Most multipath flowmeter users select models with four to eight different paths to determine mass flow. Since 1998, when the AGA approved the use of multipath ultrasonic flowmeters for custody transfer of natural gas, and other international approval agencies followed suit, there has been a substantial increase in the use of these flowmeters for natural gas measurement, especially for custody transfer.

The API has also approved a standard for using ultrasonic flowmeters for custody transfer of liquids in its Manual of Petroleum Measurement Standards. This API document only pertains to spool type, multipath ultrasonic flowmeters with permanently affixed acoustic transducer assemblies. Often this flowmeter type has a dedicated transducer installed at the 12 o'clock position to enable full pipe detection. While the standard applies specifically to custody transfer measurement, other acceptable applications include allocation measurement, check meter measurement, and leak detection measurement.

Transit Time Flowmeters Have Expanded Their Capability to Measure Fluids with Particles

Suppliers have made many technological improvements in ultrasonic flowmeters in the past 30 years. Improved electronic processing technology enables transit time meters to better handle fluids that are not entirely pure. This allows transit time flowmeters to be used for applications that could previously only be handled by Doppler flowmeters. Increased accuracy has led to broader use of ultrasonic meters in a wider variety of conditions.

The use of Doppler meters has declined with the improvements in transit time technology. Doppler appears to be almost exclusively a clamp-on technology.

Flow Calibration Facilities Are Now More Available

Certainty about precision is a requirement for end-users regardless of instrument type. Within the world of flowmeters, there are three instances when performance is evaluated:

First, a calibration process is ordinarily a mandatory step prior to shipping the device to the user. This traceable result ensures compliance with the purchase terms and is based on the needs of the intended application. This step may be performed by the manufacturer or a trusted and accredited third party.

Second, and not in all instances, users will include another calibration during their own commissioning routine. This calibration ensures that any incidents during the delivery have not affected the flowmeter's performance.

And third, flowmeters in service are calibrated to ensure compliance with current operational parameters, to adhere to the company's internal quality control routines or contractual requirements, as a response to diagnostic alarms, or to account for normal wear and tear.

One of the early barriers to the use of ultrasonic flowmeters for natural gas measurement was the availability of calibration facilities. Until 1999, there was no easily available calibration facility in the United States. Users wishing to have their meters calibrated had in many cases to send them to Europe. In 1999, Colorado Engineering Experimental Station Inc. (CEESI) opened its calibration facility in Iowa. This facility is now fully operational and can calibrate large meters in the 30 to 36-inch range, as well as smaller meters. Users in the United States and Canada can now avoid the cost and delay of sending their meters to Europe.

Soon after CEESI opened its doors, the TransCanada Calibrations high-pressure natural gas testing facility opened in Spring 2000 in Manitoba, Canada. This facility still services the Canadian market, although it is also convenient to many users in the northern United States and can handle ultrasonic flowmeters for pipe sizes up to 42 inches in diameter. TransCanada Calibrations holds itself to the highest standards, including the US National Institute of Standards and Technology (NIST) through measurement inter-comparison, AGA 9, ISO 9001, and ISO 17025.

Southwest Research Institute (SWRI) in Austin, Texas, is another facility capable of performing ultrasonic flowmeter calibrations, although they cannot easily calibrate ultrasonic flowmeters larger than 16–20 inches.

Ultrasonic flowmeters are increasingly the preferred substitute for turbine flowmeters in large line size applications for natural gas pipeline transport, and the calibration industry is adjusting to this trend. Euroloop, a flow calibration facility in Rotterdam, the Netherlands and the world's largest test facility for high pressure and high flow gas and liquid flowmeters, can now test gas ultrasonic flowmeters from 4 to 48 inches in line size diameter (and up to 30,000 m^3/h), and liquid ultrasonic flowmeters from 4 to 50 inches (and up to 5,000 m^3/h). These capabilities are large improvements over what were originally available to ultrasonic flowmeter owners. (Euroloop was initially operated by the Netherlands Metrological Institute (NMi). It later become independent and, since February 2017, has been run by Kiwa.)

One interesting Euroloop test process was developed for KROHNE's largest diameter inline ultrasonic flowmeter, the 7-path ALTOSONIC V. These flowmeters were 48 inches in diameter and the intended application called for a meter with a full-bore design to minimize pressure loss. They also required the ability to measure bidirectional flows, as some meters were to be used in loading/unloading situations. Other meters were to be used in pipeline transport, so measurement accuracy across the carrying capacity range and an ability to ascertain pipeline integrity were essential. Ultrasonic technology met all of these conditions. This 2019 Euroloop calibration used hydrocarbon gases and determined that an accuracy level of ±0.15 percent of the flowrate could be sustained over the full flow range with a certified uncertainty of 0.027 percent at ≤15,000 m^3/h. Again, ultrasonic flowmeters could not have handled these measurement parameters years ago when they were first introduced.

Cooperation has played an important role in Europe's development of new, practical standards of measurement. In the 1970s there was a general realization that new calibration standards must be developed to handle the emerging network of the natural gas grid. Ever more points of custody transfer were being installed, which meant that there was an increasing demand for reliable and stable reference values for high-pressure gas flow measurements. Of particular importance to this fast-growing energy source was the principle of third-party access in support of the direct invoicing of natural gas transport. It was rightly determined that natural gas transport entities must have a certainty in knowledge about the contents of their individual segments of the transport grid.

This need for certainty prompted several calibration groups to join in developing relevant new standards. Most important were the contributions of the Netherlands' VSL (the Dutch Metrology Institute), Germany's PTB (Physikalisch Technische Bundesanstalt), France's LNE Laboratory (now called the National Metrology and Testing Laboratory), and Denmark's FORCE Technology laboratories. Each of these groups had extensive experience in testing flow parameters and working with energy and process partners to optimize gas flow measurement technologies and equipment.

This effort to attain a single equivalent reference value for the measurement of natural gas by ultrasonic flowmeters has proven to be extremely beneficial to this technology's adoption. With a single standard agreed upon, Europe's network of test and calibration facilities was then made available to satisfy the needs of ultrasonic flowmeter suppliers and end-users alike. This level of cooperation eventually led to VSL commissioning and launching the world's first LNG calibration and test facility in July 2020 at the Maasvlakte in the Rotterdam harbor area. This facility can test and calibrate all type of flowmeters, including ultrasonic. It simulates real-life, field-scale flow and composition data at cryogenic conditions to ensure that flowmeters will work under actual operational conditions.

Many calibration facilities offer a range of services related to fluid measurement, including flowmeter diagnostics and evaluations, meter repair, witness inspection services and documentation performed to guarantee calibration and testing quality, and factory-designed training programs conducted in the customer's site or in the facility itself.

Another historic event occurred in 2011 when Emerson Process Management (now Emerson Automation Solutions) opened the first internationally certified flow calibration facility in the Middle East. Located in Abu Dhabi, the facility was in the works for about two years. In the past, customers in the Middle East sent their flowmeters to calibration facilities in the United States, Canada, or Europe for calibration. Emerson's new flow laboratory gives customers the option of reduced shipping costs and quicker turnaround time – especially important when flowmeters have to be pulled out of service for calibration, as not all companies have backup flowmeters at the ready.

Although Emerson's flow laboratory fills many of the needs of companies in the Middle East and Africa, it does have serious limitations. The facility is water-based, and for liquids only. Furthermore, and most problematical for ultrasonic technology users, is that it only handles Coriolis, magnetic, and vortex flowmeters.

So, while calibration facilities are more widely available for inline ultrasonic flowmeters, and many new areas of expertise have been established, there is still room for growth to support this measurement technology. This growth will, in part, be assisted by international accreditation agencies such as the American Association for Laboratory Accreditation, the International Laboratory Accreditation Cooperation, third-party service providers, and the further ambitions of manufacturers themselves.

Flow Measurement Redundancy Is Now More Important

Redundancy is becoming increasingly important in flow measurement. Some companies run two flowmeters in series for highly important custody transfer measurements. Some environmental reporting requirements, for example, are such that companies cannot afford to have their flowmeters go out of commission and need to build redundancy into their measurements. Redundancy is the motivation behind the development of Emerson Daniel's 3416 4+2 ultrasonic flowmeter, which uses two transmitters.

One of the newest meter configurations on the market is a "dual transmitter" type that offers a level of redundancy and reliability not previously available on a standard basis. These meters were developed for custody transfer of petroleum liquids as well as for custody transfer of natural gas, two important revenue generators for suppliers and users alike. The Daniel Dual-Configuration (Model 3410) inline flowmeter, for example, includes two fiscal meters in a single body that reduces installation costs and extends the recalibration period.

While accuracy and reliability are often cited as the two most important criteria for successful flow measurement, redundancy is fast becoming a third extremely important factor. Redundant measurement greatly increases the probability that no measurement will be lost, and that accurate measurement will continue even when one sensor or transmitter is down.

Clamp-on Models Hit Their Stride in Very Large Pipe Diameters

Clamp-on ultrasonic flowmeters are manufactured to accommodate line sizes ranging from ½ to over 230 inches in diameter. One value here is that where the process may use several different pipe diameters and carry different fluids in each, the exact same volumetric flow measurement technology can potentially be used for all points in the process. While this circumstance may not be typical, the potential reduction in employee training, spare parts, network customization, and other cost areas would be significant. Furthermore, because portable handheld clamp-on flowmeter technology exists where only the transducers must be fixed in place, the savings in avoiding the purchase of redundant equipment is multiplied.

But the primary and more certain savings is in the initial expenditure. Clamp-on designs are, essentially, immune to price variations based upon pipe size diameter. This factor must be compared to inline models of any other technology where line size can be the predominant cost factor, or where certain line sizes cannot be accommodated at all (e.g., Coriolis flowmeters have a maximum line size capability of 16 inches). Where the other components of the application's measurement requirements permit, clamp-on technology represents an excellent solution.

The Worldwide Business of Regulating Greenhouse Gas Emissions

The sun emits many types of electromagnetic radiation, including visible light, ultraviolet light, and infrared rays. Some of the infrared radiation (heat) that the sun sends our way is normally reflected back into outer space or absorbed and then released back into space at night. However, some gases absorb infrared radiation and prevent it from escaping beyond our atmosphere. The more of these gases there are in the atmosphere, the more heat is kept from escaping from earth. These gases are called greenhouse gases because they trap infrared radiation like the glass walls of a greenhouse – enough of the sun's energy is gained to build up heat inside, which can be regulated by opening windows, but the glass slows the loss enough that the inside of the greenhouse stays warmer. In a similar fashion, but without the windows option, greenhouse gases in earth's atmosphere trap the sun's infrared radiation, resulting in global warming.

One of the first international efforts to address the looming climate change issue and its causes was the Kyoto Protocol. Adopted in 1997 and ratified by its 55 signatory countries in 2005, the protocol identified six of the main greenhouse gases of most concern:

- Carbon dioxide (CO_2)
- Nitrous oxide (N_2O)
- Methane (CH_4)
- Perfluorocarbons (PFCs)
- Hydrofluorocarbons (HFCs)
- Sulfur hexafluoride (SF_6)

In the United States, the Environmental Protection Agency (EPA) originally addressed this issue through the Clean Air Act of 1990. This initiated continuous emissions monitoring (CEM) of harmful gases being emitted into the atmosphere. Flowmeter manufacturers responded with insertion ultrasonic flowmeters, differential pressure flowmeters with averaging pitot tubes, and thermal flowmeters designed to measure the volume of gases being emitted. Other instruments came onto the market in large quantities that analyzed the contents of these emissions. Europe initiated the European Union (EU) Emission Trading Scheme (ETS) to regulate the six gases listed above, with the main focus on carbon dioxide.

Beginning in 2008, the Obama administration placed a renewed emphasis on identifying and curtailing the emission of greenhouse gases and set ambitious goals for making a move from fossil fuels to renewable energy. The Trump administration reversed some of these initiatives, but the Biden administration is reviewing those rollbacks and setting ambitious goals to combat climate change.

However, the need for flare gas and stack gas measurement is not limited to the United States. It is worldwide. Expect the demand for insertion ultrasonic flowmeters used for this purpose to continue to grow worldwide as the impact of climate change becomes ever more apparent and the need for immediate remedies more pressing.

FRONTIERS OF RESEARCH

Technological Improvements over the Last Two Decades Have Continued to Expand Ultrasonic Meter Use

Advances in transit time technology in particular have broadened the types of liquids that can be measured. Because of improvements in electronic processing technology, transit time meters are better able to handle fluids that are not completely clean. Transit time flowmeters can now be used for applications that could previously only be handled by Doppler flowmeters. These improvements have also increased the accuracy of transit time ultrasonic meters, which has led to a broader use in a wider variety of conditions. Expect suppliers to continue this research as they seek to measure the flow of less clean fluids and also seek to increase the accuracy of transit time meters.

More Research into Doppler Meters

Even though suppliers of transit time meters have cut into the market share of Doppler meters, there is still and opportunity for Doppler meter suppliers to increase the performance of their meters. Doppler meters have been losing market share to transit time suppliers over the past five years. Their main competition is from magnetic flowmeters, especially for wastewater applications and other dirty liquid applications like pulp and paper. The increased need to measure water and wastewater, along with open channel flow, presents an opportunity for Doppler suppliers.

More Research into Custody Transfer Applications

Ultrasonic meters shine in custody transfer applications, especially for custody transfer of natural gas. Suppliers are bringing more of these meters onto the market. As part of this effort, companies will continue to experiment with transit time meters with more paths. Some of these paths are devoted to self-diagnostics. Baker Hughes and Honeywell Elster are two of the most technologically driven ultrasonic suppliers, along with SICK. Developing meters for the growing LNG measurement market is part of this total picture. Baker Hughes has an ultrasonic meter designed for LNG measurement.

Portions of the discussion of chords and paths in this chapter are excerpted from an article written by the author and published in the September 2012 issue of *Flow Control*. The article is entitled “The Path to Chordal Harmony.”

7 Vortex Flowmeters

OVERVIEW

Vortex flowmeters were first introduced by a company called Eastech in 1969. It was Yokogawa that popularized this flowmeter in the early 1970s. Since that time, other major suppliers have come into the market, including Endress+Hauser and Emerson Rosemount. Other companies with a significant market presence in vortex include ABB, Foxboro, and VorTek Instruments.

Like many other flowmeters, vortex meters experienced application issues early in their development. One problem for vortex meters was the effects of vibration, which tended to create false readings. Suppliers created software to deal with vibration issues, and this proved to be an effective solution.

Vortex flowmeters are different from most other new-technology flowmeters in that they penetrate the flowstream, although in a limited way. Magnetic and ultrasonic flowmeters have very little impact on the flowstream. However, vortex meters operate by means of a bluff body suspended in the flowstream. A bluff body is a broad, flat object, often made of metal. A bluff body in the flowstream generates vortices, which are detected by a sensor on one or both sides of the pipe. The flowmeter counts the number of vortices, and this number is proportional to flowrate.

Vortex meters are among the most versatile type of meter. They can readily measure liquid, gas, and steam flow. Other new-technology meters such as ultrasonic and Coriolis can readily measure liquids and gases but have difficulty with steam. Thermal meters are used almost exclusively for gas flow measurement. Magnetic flowmeters excel at measuring liquids but cannot measure steam or gas flow. Vortex meters are the only new-technology meter that readily measures all three fluid types.

Vortex flowmeters are well suited for measuring steam flow, and they are widely used for this purpose. Steam is the most difficult fluid to measure. This is due to the high pressure and high temperature of steam, and because the measurement parameters vary with the type of steam. Main types of steam include wet steam, saturated steam, and superheated steam. Steam is often measured in process plants, and for power generation. In addition to their ability to tolerate high process temperatures and pressures, vortex meters have wide rangeability. This allows them to measure steam flow at varying velocities. In process and power plants, steam is often measured coming from a boiler.

MULTIVARIABLE FLOWMETERS

Multivariable flowmeters measure more than one process variable. In addition to volumetric flow, they typically have a pressure transmitter and a temperature

DOI: 10.1201/9781003130017-7

sensor and/or transmitter. These values are used to compute mass flow. In the past, these values were measured independently and then output to a flow computer, which performed the mass flow calculation. Now the computation for mass flow is being done within the flowmeter itself, and the volumetric flow, pressure, and temperature values are fed into the multivariable flowmeter for calculation of mass flow. Multivariable vortex flowmeters can also readily measure steam flow.

Cost is one advantage of multivariable vortex meters. While they cost more than single variable meters, they typically cost less than buying the components separately. These typically include a single variable vortex meter, along with a pressure and temperature transmitter. In addition, multivariable vortex meters have the advantage that the components can be calibrated together as one unit before being placed in the field, giving higher reliability and potentially higher accuracy.

VORTEX FLOWMETERS PROVIDE ACCURATE AND RELIABLE FLOW MEASUREMENT AT A COMPETITIVE PRICE

It is difficult to describe vortex meters in a few words. They have some disadvantages when compared to other new technology flowmeters. Vortex meters are not as accurate as Coriolis meters, they are more intrusive than ultrasonic meters, and are less widely used than magnetic meters. They are more expensive than differential pressure (DP) flowmeters. Yet despite these comparisons, vortex flowmeters offer a number of important advantages when compared to both new technology and traditional technology meters.

Vortex flowmeters offer accurate and reliable flow measurement at a competitive price. Even though vortex meters are not as accurate as Coriolis meters, many vortex meters offer accuracy readings of better than one percent, depending on fluid and application. While vortex meters are somewhat intrusive compared to magnetic and ultrasonic meters, they are much less intrusive than the orifice plates used with DP transmitters to measure flow. Pressure drop from vortex meters is minimal, since most shedder bars are relatively small in size. While vortex meters are less widely used than magnetic flowmeters, this is due in part to the fact that they were introduced nearly 20 years after magmeters.

Vortex flowmeters offer an accurate and reliable means to measure flow. They cannot achieve the high accuracy of Coriolis flowmeters, but neither can any other type of flow measurement device. The most common applications do not require the absolute highest accuracy available and can be suitably addressed with that offered by vortex flowmeters. Vortex measurement accuracy can be as high as ±0.65 percent for liquids and ±0.9 percent for gases and steam, with repeatability of ±0.1 percent of rate for liquids, gases, or steam.

Another strength of vortex meters is that they can handle a wider range of process conditions than almost any other flowmeter. Vortex meters can readily handle liquids, gases, and steams. While Coriolis meters are well known for their ability to handle liquids, they are less widely used on gases, and their use on steam is very recent. Ultrasonic meters were used in the 1980s to measure gas flow but are only beginning to be used for steam. By contrast, the use of vortex meters to measure steam flow is well established. Vortex meters can handle the high pressures and temperatures associated with steam flow measurement.

THE ORIGIN OF MULTIVARIABLE VORTEX FLOWMETERS

Vortex flowmeters display a growing trend toward multi-technology flowmeters. Single variable vortex meters measure volumetric flow. Multivariable vortex flowmeters include pressure and/or temperature sensors for the purpose of measuring mass flow. By using the information from these sensors, the flowmeter can determine volumetric flow, temperature, pressure, fluid density, and mass flow. This multivariable flowmeter is one of a growing number of multivariable new-technology flowmeters, including multivariable magnetic flowmeters and multivariable DP flowmeters. Multivariable ultrasonic flowmeters are popular in district heating applications.

The origin of the multivariable vortex meter goes back to VorTek Instruments. Jim Storer had the idea for a multivariable vortex meter in 1994 while at EMCO Flow Systems. At that time, EMCO was selling vortex flowmeters along with pressure and temperature sensors and sending the values to a flow computer to calculate mass flow. Jim Storer had the idea to consolidate all these measurements in a single unit to create a multivariable vortex meter. In 1995, he left EMCO to found VorTek Instruments in Longmont, Colorado, where he began the design of the multivariable vortex flowmeter.

Storer lacked a distribution channel and for this purpose, he formed a joint venture with Sierra Instruments in Monterey, California. According to the terms of the agreement, VorTek retained the rights to and would manufacture the insertion multivariable vortex meters, while Sierra would own and manufacture the inline version. However, Sierra would be allowed to market both meters under its brand name. This arrangement lasted for several years until VorTek developed a second generation multivariable vortex flowmeter that Sierra sold under its brand name. This arrangement continued until about 2014, when Sierra decided to manufacture its own vortex meter. Even after this, Sierra continued to sell VorTek meters under the Sierra name. This arrangement continues to this day. In 2017, Jim Storer retired from VorTek and turned the reins of the company over to Eric Sanford, who still serves as president.

Since multivariable vortex meters were first introduced, a number of other companies have come out with their own multivariable vortex flowmeters. These include ABB, Yokogawa, KROHNE, and Endress+Hauser. While multivariable flowmeters are somewhat more expensive than their single-variable counterparts, they enable users to obtain significantly more information about the process than a single-variable volumetric meter. This additional information can result in increased efficiencies that more than make up for the additional cost of the

TABLE 7.1
Advantages and Disadvantages of Vortex Flowmeters

Advantages	Disadvantages
Extremely versatile – measure liquid, steam, and gas	May cause some pressure drop due to presence of bluff body
Relatively high accuracy	
Have no moving parts	Very limited use for custody transfer
Multivariable meters provide mass flow measurement	Measurement is affected by pulsating flow
Especially well-suited for steam flow measurement	Difficulty measuring slurries or high viscosity flow
Dual sensors provide redundancy	
Standards have been developed for use in custody transfer applications by the American Petroleum Institute	Still need additional industry approvals for custody transfer measurement

multivariable flowmeter. Multivariable vortex flowmeters also have the capability of measuring mass flow, and this makes them attractive, especially for steam and gas flow measurement.

Multivariable flowmeters measure more than one process variable. In addition to process fluid flow, multivariable vortex flowmeters have one of the following three sensor configurations:

- Integrated temperature sensor; no pressure sensor
- Integrated temperature and pressure sensors
- Integrated pressure sensor; no temperature sensor

About ⅔ of multivariable vortex flowmeters are shipped with an integrated temperature sensor but with no pressure sensor (Table 7.1).

VORTEX FLOWMETER COMPANIES

VorTek Instruments

VorTek Instruments, an azbil Group Company, manufactures precision multivariable flowmeters liquid, gas, steam, and district energy. Since January 2013, VorTek Instruments has been a subsidiary of Azbil North America, a wholly owned subsidiary of the Japan-based azbil Group.

History and Organization

VorTek Instruments was founded in 1995 as a high-quality manufacturer of flow measurement instruments. The company has specialized in vortex technology, and during its history has routinely designed and manufactured custom flowmeters to meet its customers' original equipment manufacturer (OEM), private label, or general customer needs and specialized application requirements. This practice continues today.

Vortex Flowmeters

VorTek's multivariable vortex meters incorporate a high-accuracy velocity sensor, a precision platinum RTD temperature sensor, and a solid-state pressure transducer to measure the mass flowrate of gases, liquids and steam. The Pro-V™ Model M24 can deliver volumetric flow, mass flow, temperature, pressure, density, and energy (BTU) measurements from a single installed device.

YOKOGAWA ELECTRIC CORPORATION

Yokogawa Electric Corporation is a leading provider of industrial automation and test and measurement solutions, with 119 companies and operations in 61 countries. The company develops, manufactures, and markets information technology (IT) solutions, measuring and control equipment, semiconductors, and electronic components. Yokogawa's products include pressure transmitters, flowmeters, analyzers, data recorders, IT controllers, switching power supplies, and AC (alternating current) adaptors. In flow measurement, Yokogawa is the leading supplier of vortex flowmeters worldwide, and also provides magnetic, variable area, Coriolis, and differential pressure flowmeters.

Tamisuke Yokogawa, Doctor of Architectural Engineering, established an electric meter research institute in Tokyo in 1915, and Yokogawa became the first company to both produce and sell electric meters in Japan. Ichiro Yokogawa and Susumu Aoki established Yokogawa Corporation in December 1920. Recent executive appointments include the naming of Takashi Nishijima to serve as Chairman, and Hitoshi Nara to serve as President and Chief Executive Officer. Yokogawa Corporation of America was established in 1957.

Yokogawa's industrial automation (IA) and control business – home of the company's flowmeters, pressure transmitters, and other field instruments – is by far the largest contributor to Yokogawa's bottom line, generating more than 90 percent of corporate revenue. This business is now optimizing solutions by focusing on three industry segments: energy and sustainability, materials, and life.

In addition to the industrial automation and control business, Yokogawa is developing the following two independent businesses: Measuring Instruments Business (high-precision technology for environment-related electrical power and optical communications) and New Businesses and Other (IoT hardware, software, and cloud solutions for service providers and an aircraft instruments business).

Vortex Flowmeters

Yokogawa's digitalYEWFLO variable meters measure steam, gases, and liquids, including saturated steam and high temperature oil, and LNG. The company emphasizes its field-proven sensor technology and high reliability, as well as four-wire and two-wire designs. The digitalYEWFLO is claimed to be the world's first two-wire multivariable type (with built-in temperature sensor) that can directly output the mass flowrate of saturated steam. There are also four-wire designs.

HOW THEY WORK

Vortex flowmeters operate on a principle called the von Karman effect. This principle concerns the behavior of fluids when an obstacle is placed in the path of flow. Under the right conditions, the presence of the obstacle generates a series of alternative vortices called the von Karman street. This phenomenon occurs in liquid, gas, and steam and has been observed in many diverse contexts, including cloud layers passing an island and whitewater rapids.

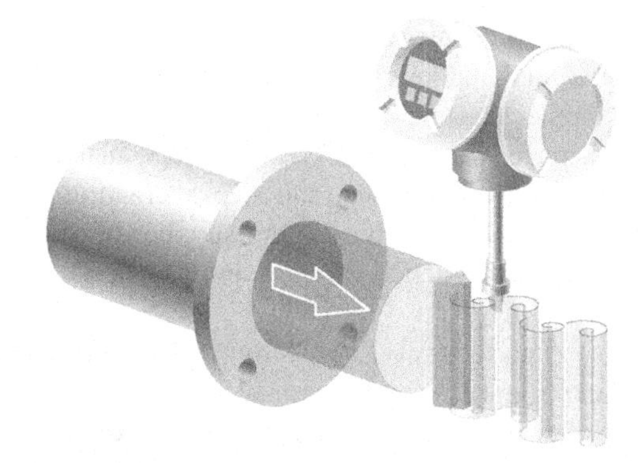

The principle of operation of a vortex flowmeter. (Source: Courtesy Endress+Hauser)

In vortex flowmeters, the obstacle takes the form of an object with a broad, flat front called a bluff body that is mounted at right angles to the flowstream. Flow velocity is proportional to the frequency of the vortices and flowrate is calculated by multiplying the area of the pipe times the velocity of the flow.

In order to compute the flowrate, vortex flowmeters count the number of vortices generated by the bluff body using a variety of techniques. The majority of vortex flowmeters use a piezoelectric sensor to detect vortices; however, some use a capacitive or ultrasonic sensor.

Vortex Flowmeter Performance

The vortex flowmeter market has not grown as rapidly as the Coriolis and ultrasonic markets, and it is not nearly as large as the magnetic flowmeter market. One factor holding it back is that it is more intrusive than those other three meters. A bluff body can get knocked out of place, as can an orifice plate, and the result is an unreliable measurement. Coriolis, ultrasonic, and magnetic flowmeters are not intrusive in this way, and they provide more stable measurements.

Secondly, there is a limit to how much information the vortex transmitter can derive from counting the number of vortices. Apparently this is not sufficient for vortex flowmeters to achieve accuracies of much over 0.5 percent. Ultrasonic and Coriolis meters reach higher levels of accuracy. In order to achieve a breakthrough in accuracy or performance, someone will have to figure out how to derive additional information from the vortices than is currently being derived.

Some companies have taken steps to improve the capability of their flowmeters at low flowrates by using a very small shedder bar that is more sensitive to lower flowrates than meters with larger shedder bars. While this technology can push back the limits of low flow, it does not completely solve the problem of low flow measurement.

Another approach to handling low flowrates is through the use of reduced bore meters. Reduced bore vortex flowmeters are narrower in diameter where the bluff body generates vortices than at either end of the flowmeter. Generally, the internal diameter of the meter is reduced by either one meter size or two meter sizes. Flow speed is increased where the pipe is narrowed. Reduced bore vortex flowmeters can more easily handle low flowrates. Reduced bore meters are also typically easier to install than standard bore meters, and their installation costs tend to be less.

While reduced bore vortex flowmeters have only appeared fairly recently, beginning with the single line size versions, they are becoming increasingly popular. Look for more suppliers to introduce them in the next several years, including both one-meter and two-meter size reduced bore meters. Their use will increase as more customers learn about their advantages and have a chance to try them out on their own applications. Emerson Rosemount and Yokogawa are among the companies that manufacture reduced bore vortex meters.

GROWTH FACTORS FOR THE VORTEX FLOWMETER MARKET

Growth factors include:

- Steam flow is a growing application for vortex flowmeters.
- Product consolidation and new suppliers have changed the market dynamics.
- More companies are introducing multivariable flowmeters.
- Custody transfer of natural gas.
- Dual sensor designs.
- District heating applications are becoming more popular.

Steam Flow Is a Growing Application for Vortex Flowmeters

Vortex flowmeters are well suited for measuring steam flow, and they are widely used for this purpose. Steam is the most difficult fluid to measure. This is due to the high pressure and high temperature of steam, and because the measurement parameters vary with the type of steam. The main types of steam are wet steam, saturated steam, and superheated steam.

Steam is often measured in process plants, and for power generation. Vortex and DP flowmeters are currently the primary flowmeters used to measure steam. Magnetic flowmeters cannot measure steam flow, and Coriolis and ultrasonic flowmeters have not significantly entered this market due to their own technical limitations.

These factors leave the door open to vortex flowmeters to maximize their value in this area with less competitive intrusions than in most other application sets.

In addition to their ability to tolerate high process temperatures and pressures, vortex meters have wide rangeability. This allows them to measure steam flow at varying velocities. In process and power plants, steam is often measured as it comes directly from a boiler as well as further downstream from the source, making the adaptability of the vortex measurement a real plus.

Product Consolidation and New Suppliers Have Changed the Market Dynamics

Probably the most important change to the vortex flowmeter market in the past 25 years occurred in 1994 when Emerson Rosemount entered the market. Emerson Rosemount became one of the few US-based suppliers of vortex flowmeters. Because Emerson Rosemount is a broad-line supplier of flowmeters, they are able to offer users a wide range of choices in flowmeters. The company also brought news and information about vortex meters to a whole new group of customers. As a result, Emerson Rosemount has become the market share leader of the North American market.

In the past ten years, some new companies have entered the multivariable vortex flowmeter market. These instruments can output mass or volumetric flow, pressure, temperature, density, or concentration information. Engineering designs that minimize or eliminate process vibration have also come on to the market and improved both the perceived and actual accuracy of the base technology. The presence of these new suppliers and enhanced product lines inject new life into the overall flowmeter market and is helping to spread vortex technology to a still larger set of customers.

Some of the new companies that entered the multivariable vortex flowmeter market previously had only single-variable vortex flowmeters. In addition to this, some new companies have entered the vortex market for the first time or have re-entered the market. These companies include Aalborg, azbil, and Nice Instrumentation. Racine Federated (now part of Badger Meter) purchased the industrial vortex flowmeter product line from J-TEC Associates. Spirax Sarco purchased the vortex flowmeter line from EMCO Flow Systems, although they have now discontinued manufacturing this line. Universal Flow Monitors brought out a new line of plastic vortex flowmeters called CoolPoint. The presence of these new suppliers and product lines has injected new dynamics into this flowmeter market and has helped to spread this technology to still more customers.

More Companies Are Introducing Multivariable Flowmeters

Sierra Instruments introduced the first multivariable vortex flowmeter in 1997. This meter included an RTD temperature sensor and a pressure transducer with a vortex shedding flowmeter. By using the information from these sensors, the flowmeter can determine volumetric flow, temperature, pressure, fluid density, and mass flow. This multivariable flowmeter is one of a growing number of multivariable new technology flowmeters, including multivariable magnetic flowmeters and multivariable DP flowmeters. Multivariable ultrasonic flowmeters are popular in district heating applications.

A number of other suppliers have since introduced multivariable vortex flowmeters. These include ABB, Yokogawa, KROHNE, VorTek (now an azbil subsidiary), and Endress+Hauser. While multivariable flowmeters are somewhat more expensive than their single-variable counterparts, from a single entry point in the pipeline they can offer precise measurements of mass or volumetric flowrates – a choice that single variable flowmeters cannot offer. And multivariable functionality from a single meter helps to ensure optimal cost savings and simpler installations. This additional information can result in increased efficiencies that more than make up for the additional cost of the multivariable flowmeter.

Multivariable vortex flowmeters also have the capability of measuring mass flow, and this makes them attractive, especially for the higher growth areas of steam and gas flow measurement.

Custody Transfer of Natural Gas

Custody transfer of natural gas is a fast-growing market, especially with the increased popularity of natural gas as an energy source. Natural gas changes hands, or ownership, at a number of points between the producer and the end-user. This changing of hands occurs at custody transfer points, and it is an application that is highly regulated by the American Gas Association (AGA), the American Petroleum Institute (API), and other approval authorities internationally. In January 2007, an API committee approved a draft standard for the use of vortex flowmeters for custody transfer of liquid and gas. This approval has provided some boost in sales of vortex meters over time, as suppliers develop products that conform to the standard. The standard was first updated in 2010 and then reconsidered in 2013. Eventually, it was adopted as a custody transfer standard for steam and gas.

Dual Sensor Designs

Another recent development in vortex flowmeters is dual sensor meters. When a bluff body generates vortices, a downstream sensor detects the vortices and counts how many there are. These sensors are typically ultrasonic, piezoelectric, or capacitive, though piezoelectric sensors predominate. One recent innovation is to have two sensors downstream to detect the vortices. This provides redundancy, making the flowmeter more reliable, though it doesn't necessarily enhance performance.

Another innovation involving multiple flowmeters is to have two vortex flowmeters in the same process line and calibrate them together. This also provides redundancy. It is somewhat like having two ultrasonic flowmeters running in the same line for custody transfer applications. This arrangement does not improve individual meter performance, but it does make for a more stable and reliable installation.

Vortex flowmeters with dual sensors downstream and dual vortex flowmeters calibrated together each account for less than five percent of the vortex flowmeters shipped. However, the market for both configuration types can be expected to grow at an above average rate over the next several years. As more end-users become familiar with the advantages of redundancy that both designs offer, their popularity can be expected to increase.

District Heating Applications Are Becoming More Popular

District heating is a growing application for vortex flowmeters. In district heating, a centralized heating system is created that can provide heat for a building, a group of houses, or a campus. A similar concept applies to district cooling. In district heating, a volumetric flowmeter is often combined with a temperature sensor and pressure transmitter to measure mass flow. A flow computer is used to calculate mass flow. Both vortex and ultrasonic flowmeters are used to measure water flow, and steam flow is also measured.

District heating is a growing application in Europe, and it is also beginning to be more widely used in Asia. It has never caught on to the same degree in North America, although there is measurably more interest in the application in this region. It is likely that district heating will start being used more widely in the United States as the costs of energy increase and heating providers and consumers alike seek more control. District heating is already more widely used in Canada.

Measuring Low Flowrates

One weakness of vortex technology is the difficulty that vortex flowmeters have in measuring low flowrates. This difficulty is inherent in the technology itself. Shedder bars may not shed vortices at all, or they may not shed vortices at a regular rate if the flow is too low. Other meters, such as thermal mass meters, do a better job than vortex meters of measuring flow at low flowrates.

In a no-flow situation, a vortex meter will register zero flow. However, it is unable to immediately respond as flow begins. A vortex meter with an output of 4–20 mA will have an output of 4 mA at zero flow. As flow picks up, it may suddenly jump to 4.8 mA when the flowrate is fast enough to register on the vortex meter. The flowrate that is represented by the readings between 4 mA and 4.8 mA is a "blind spot" for the meter.

Some companies have taken steps to improve the capability of their flowmeters at low flowrates by using a very small shedder bar that is more sensitive to lower flowrates than meters with larger shedder bars. While this technology can push back the limits of low flow, it does not completely solve the problem of low flow measurement.

Another approach to handling low flowrates is through the use of reduced bore meters. Reduced bore vortex flowmeters are narrower in diameter where the bluff body generates vortices than at either end of the flowmeter. Generally, the internal diameter of the meter is reduced by either one meter size or two meter sizes. Flow speed is increased where the pipe is narrowed. Reduced bore vortex flowmeters can more easily handle low flowrates. Reduced bore meters are also typically easier to install than standard bore meters, and their installation costs tend to be less.

While reduced bore vortex flowmeters have only appeared fairly recently, beginning with the single line size versions, they are becoming increasingly popular. Look for more suppliers to introduce them in the next several years, including both one-meter and two-meter size reduced bore meters. Their use will increase as more customers learn about their advantages and have a chance to try them out on their own applications. Emerson Rosemount and Yokogawa are among the companies that manufacture reduced bore vortex meters.

FRONTIERS OF RESEARCH

Multivariable Flowmeters

Multivariable vortex flowmeters were first invented by Jim Storer, founder of VorTek Instruments, in 1996. Multivariable flowmeters measure volumetric flow, along with pressure and temperature, and use those three values to compute mass flow. This is called an inferred or indirect method of determining mass flow. Storer founded Vortek initially to manufacture these multivariable vortex meters. As explained above, he made a marketing agreement with Sierra Instruments for Sierra to sell these meters under the Sierra brand name. Since that time, quite a few more companies have brought out their own multivariable vortex meters. Expect existing suppliers to work to enhance performance and expect to see more suppliers enter this market.

Dual Meters

VorTek Instruments has introduced the VorCone meter, which combines a vortex meter with a cone meter. So far the reception has been quite positive, according to Eric Sanford, president of VorTek. There are similar dual meters involving other technologies. For example, some companies market a meter that combines a positive displacement and a turbine meter. This meter is used in apartment buildings. The positive displacement meter measures flow at night when water use is very low, and the turbine meter measures water flow in the morning and evening, when water use is much higher. These meters are called compound meters. Another example of this type is the Accelabar by Armstrong Veris, which combines a flow nozzle with a pitot tube.

Expect to see more of these combination meters as researchers think "out of the box" and attempt to solve new flowmetering challenges.

Dual Tube Meter

The author of this book has two patents on a dual tube meter. The idea behind the patents is to make the flow measurement using two sensors, both of which are placed inside the larger meter body. The two sensors measure flow simultaneously, and the readings go to a transmitter that uses both inputs to come up with a single reading. One advantage of this meter is redundancy, since if one sensor fails, the other one will continue to provide a reading. The original idea behind the meter was that it may be more efficient and less costly to make two independent flow readings inside a large pipe, rather than relying on just one sensor to make a large pipe reading. VorTek Instruments has built a prototype of a vortex dual tube meter, and it has been tested at Colorado Engineering Experiment Station, Inc. (CEESI). Hoffer Flow Controls has also built a turbine prototype of this meter and it has also been tested on the oil stand at CEESI.

Steam Flow Continues to be a Strength for Vortex Meters

Vortex are among the most versatile meters; they can measure steam, liquid, and gas with equal ease. The major competitors to vortex meters in steam flow are differential pressure (DP) meters. Expect suppliers to continue to study the best ways to measure saturated, wet, and supersaturated steam. After an effort that began in 2007, the American Petroleum Institute (API) approved vortex meters for custody transfer of steam and gas flow in 2017.

8 Thermal Flowmeters

OVERVIEW

Thermal flowmeters are among a small group of fluid measurement technologies that can generate a mass flow measurement independent of additional component technologies. Another flowmeter technology with this capability is Coriolis. Mass flow data is generally considered more useful than volumetric data for many applications – such as custody transfer of hydrocarbon fluids – and can offer higher repeatability, a level of measurement certainty highly valued by process and quality control managers alike.

Thermal flowmeters measure mass flow quite differently from Coriolis flowmeters. For instance, instead of using fluid momentum as Coriolis flowmeters do, thermal flowmeters make use of the thermal or heat conducting properties of fluids to determine mass flow. While most thermal flowmeters are used to measure gas flow, some also measure liquid flow.

In contrast, other flowmeter types that can also produce a mass flow result do so with the addition of pressure and temperature sensors. Mass flow is then computed from the volumetric flow, and the temperature and pressure variables. Multivariable differential pressure and vortex flowmeters can both be used in this way to compute mass flow.

THE ORIGINS OF THERMAL FLOWMETERS

Much of the original research that resulted in today's thermal mass flowmeters was conducted in the first two decades (1900–1920) of the 20th century. Hot-wire anemometers were the precursors of what we know today as thermal flowmeters. The basic concept of these anemometers was to introduce a heated element into a flowstream, to allow this element to be cooled by the incident flowing medium, and then from the temperature (or resistance) attained by the element to derive flowrate information. The use of two elements – or sensors – properly placed permitted the effects of turbulent flows to be better taken into account.

These hot wire anemometers were originally used for airflow measurement and were useful in investigations of velocity profile and turbulence research. The anemometers were very small and fragile, consisting of a thin wire element. This construction produced relatively fast response times, but their fragility made them unsuitable for typical industrial environments.

Thermal flowmeters as we know today were introduced for industrial applications in the 1970s. Credit for their introduction to the measurement market is generally attributed to three companies. These are Fluid Components International (FCI), Kurz Instruments, and Sierra Instruments. Sierra Instruments and Kurz

DOI: 10.1201/9781003130017-8

approached the subject through hot wire anemometers. FCI approached the subject through flow switches. All three companies were pioneers in the development of thermal flowmeters, and all three companies still offer enhanced versions of these thermal flowmeters today.

Hot wire anemometers consist of a heated, thin wire element, and are very small and fragile. Hot wire anemometers are used in velocity profile and turbulence research. Because they are susceptible to breakage and to dirt, they are not suited to industrial environments. Thermal flowmeters were invented by taking the principle of operation of hot wire anemometers and modifying it to make the anemometers better suited to an industrial environment. Industrial thermal flowmeters use the anemometer concept of measuring the speed of heat dissipation to determine mass flow but introduce more rugged sensors that are better adapted to industrial environments.

DEVELOPMENT HISTORY OF THERMAL FLOWMETERS

Dr. Jerry Kurz and Dr. John Olin are two of the founders of thermal flowmeter technology. This technology grew out of research they both were doing on hot wire anemometers at Thermo-Systems Engineering Co., now TSI, in Minnesota. In an interview the author conducted with Dr. Olin, he described hot wire anemometers this way:

> A hot-wire anemometer is a research tool, and it is a thermal mass flow meter. Its sensor usually is a tungsten wire a fraction of a thousandth of an inch in diameter and is very fragile. It can get into small places. It is so small and fast in response that it can measure turbulence; so, turbulence research was normally done that way. ...
>
> I mainly worked on thermal mass flowmeters. They made a thermal-anemometer probe – a probe with a hot tungsten wire on the end of it. You can measure free air flows in a room using a probe. You take the same probe and insert it in a pipe, and now it becomes a mass thermal flow meter. The hot-wire anemometers work very much like the thermal mass flowmeters currently manufactured by Sierra, FCI, and Kurz.
>
> In the case of hot-wire anemometers, the electronics puts electrical current through the tungsten wire. The tungsten has a high resistance; so, we get self-heating ($I^2 R$). It self-heats itself above the temperature of the flowing air or gas.
>
> Most manufacturers use a constant-temperature thermal anemometer. The electronics keeps the temperature of the heated tungsten wire at a constant value above the temperature of the gas flow. So, we have to concurrently measure the temperature of the gas flow. In thermal anemometers, you have to keep something constant. It really is a constant temperature differential anemometer. It keeps the difference between the gas temperature and the heated sensor at a constant value. In the case of the constant-temperature anemometer, increasing flow increasingly cools off the heated hot wire.
>
> As the flow increases, more heat (watts) is carried away. So, it takes more current (or voltage) across that sensor to keep the temperature differential constant, i.e., it pumps more current into it. The watts required is the output of the instrument, which increases monotonically, but non-linearly, as the flow increases.
>
> In the case of FCI, they keep the electrical power constant. This is called a constant-current anemometer.

In the above discussion, Olin describes in more detail how thermal flowmeters developed out of hot-wire anemometer technology. As he tells it, hot wire anemometers were used to measure the flow of free air in a room using a probe. They were also used in turbulence research. The innovation that created thermal flowmeters came about when this probe was inserted into a pipe. However, because the tungsten wire was too fragile to measure gas flow inside a pipe in industrial applications, the probe had to be industrially hardened to withstand the environment consisting of gas flowing through a pipe.

Dr. Kurz and Dr. Olin collaborated in the early 1970s to develop thermal flowmeters as an offshoot of their research into air velocity profile and turbulence research using hot-wire anemometers. However, how they got together is a slightly convoluted story. Dr. Olin describes this story as follows:

> [In] 1971, but there was a recession in America, and I got caught up in that. So, at first TSI said, "Why don't you work in marketing?" So, I worked in marketing. I loved it. As I look back, this was great training for me when we started Sierra Instruments. Finally, I left TSI. I took a job with the Minnesota Pollution Control Agency (MPCA). I became the head of Air Quality Monitoring.
>
> TSI had two product lines: thermal anemometers and particulate instruments for the measurement of airborne particulates. I invented several particulate sampling products for TSI, especially for use in air quality monitoring. My background in particulates helped me get the job at the MPCA. I stayed there until 1974.
>
> I ultimately became the Deputy Director of the entire agency. I had air, water, solid waste, and air and noise pollution control under me. I wrote the first state implementation plan for air quality control for Minnesota. I was responsible for the first ambient air asbestos regulations. So, I had a good experience there.
>
> My boss, Director Frank Merritt, quit. So, Governor Wendell Anderson called me into his office. He asked if I wanted to take over the Agency. I said no, I would rather work for my own company – Sierra Instruments. You see, Jerry Kurz and I tried to convince TSI to make industrial products, and they said no. I got laid off, and Jerry quit TSI later. Jerry and I got together, and we said let's start our own company and make industrial thermal anemometers, as well as particulate instrumentation for the burgeoning air-pollution control market. We got the name "Sierra" from the Sierra Nevada Mountains in California.

In 1973, Kurz and Olin collaborated to form a company called Sierra Instruments. They incorporated the company in Minnesota in 1973. They developed a thermal flowmeter that was more rugged and hardened than the hot-wire anemometers. In 1975, the two entrepreneurs packed up their business into two trucks and drove it across the Continental Divide to Monterey, California. In 1977, the pair had an "industrial divorce." At that time Sierra was making both air sampling products and thermal flowmeters. Olin stayed with Sierra and kept the air sampling products. Kurz kept the thermal flowmeters and formed Kurz Instruments. Both companies are still located in Monterey, California.

Fluid Components International (FCI) took a different approach to developing thermal flowmeters. FCI was founded in 1964 by Mac McQueen and Bob Deane.

The company developed flow switches to detect the flow of oil through pipes in the oil patch. Although these thermal switches were not flowmeters, they formed the basis of what later became flowmeters. In 1981, FCI put more sophisticated electronics on its switches, creating thermal flowmeters for gas flow measurement.

THERMAL FLOWMETER COMPANIES

Fluid Components International

Fluid Components International (FCI) provides flow, liquid level, interface, and temperature sensors utilizing the company's patented thermal dispersion measurement technologies. FCI's products are used for a variety of industrial process and plant applications.

History and Organization

FCI was founded by Malcolm McQueen and Bob Deane in Canoga Park, California as a manufacturer of thermal dispersion flow switches used to detect flow movement in oil well pipes. In 1980, company facilities were moved to the present location in San Marcos, California. The company's first commercially available flowmeter was produced in 1981, when FCI added more sophisticated electronics to its flow switches. This new thermal flow product was first used in gas flow applications. The company opened a sales and service center in the Netherlands in 1993. FCI Measurement and Control Technology, a subsidiary located in Beijing, China, was opened in 2010. FCI also owns FCI Aerospace which manufactures flow, level, temperature, and pressure sensors for the aircraft industry.

In addition to its manufacturing capabilities, the company also operates its own flow and level calibration facility. The San Marcos facility has NIST traceable flow calibration stands used in both product development and customers' production product calibrations. FCI is also ISO-9001 and AS9100 certified and carries extensive certifications for the nuclear power industry.

Thermal Flowmeters

FCI designs, manufactures, and sells air and gas mass flowmeters utilizing thermal dispersion measurement technology. FCI provides several thermal mass flowmeter product families and more than 20 models, including inline and insertion configurations, to meet a broad range of performance, pipe diameters, gas types, and environmental application requirements.

FCI's ST thermal mass flowmeters for air and gas measure air, compressed air, nitrogen, hydrogen, oxygen, argon, CO_2, ozone, other inert gases, natural gas, and other hydrocarbon gases. The meters are communications friendly as they support 4–20 mA analog, frequency/pulse, or advanced digital bus communications such as HART, Foundation Fieldbus, PROFIBUS, and Modbus.

Kurz Instruments

Kurz Instruments is considered a pioneer in the development of thermal instrumentation and rugged thermal industrial devices. The company manufactures inline and insertion thermal mass flowmeters designed to monitor and measure dry air, dry gas, and wet gas environments.

History and Organization

Kurz Instruments, Inc. was founded by Dr. Jerome Kurz in 1976 after he revolutionized air flow measurement by developing a series of thermal anemometer sensors that electronically measured gas flow.

In 1986, Kurz Instruments, Inc. moved from Carmel Valley, California, its original location, to its current facilities in Monterey, California. The company experienced a leadership transition in April 2008 when Dan Kurz was appointed as president. Previous to this, he had been serving as general manager for over eight years. He is currently the CEO of Kurz Instruments.

Thermal Flowmeters

Within the thermal mass flowmeter product range, the firm offers inline, insertion, multipoint systems, and portable design types that all have full temperature range capabilities.

The inline mass flowmeters series includes the 504FTB and 534FTB. The 504FTB Series includes ten models that cover line sizes from ½ to 4 inches, and mass flow rates of up to 1,000 SCFM. The 534FTB Series can accommodate line sizes up to eight inches in diameter. Both of these series are designed for use in measuring industrial and process gas flows, compressed air, combustion gas, natural gas, and a variety of other gases.

Kurz also offers a variety of insertion thermal flowmeters.

HOW THEY WORK

There are two different thermal flowmeter technologies. Some measure the speed with which heat added to the flowstream disperses. Others measure the temperature difference between a heated sensor and the ambient flowstream. Thermal flowmeters typically require one or more temperature sensors to measure the fluid temperature at specific points (Figure 8.1).

Thermal flowmeters work by introducing heat into the flowstream and measuring how much heat dissipates, using one or more temperature sensors (Figure 8.2).

There Are Two Different Methods for Doing This

One method is called the constant temperature differential. Thermal flowmeters that use this method have two temperature sensors. One is a heated sensor, and the other sensor measures the temperature of the gas. Mass flowrate is calculated based on how much electrical power is required to maintain a constant temperature difference between the two temperature sensors.

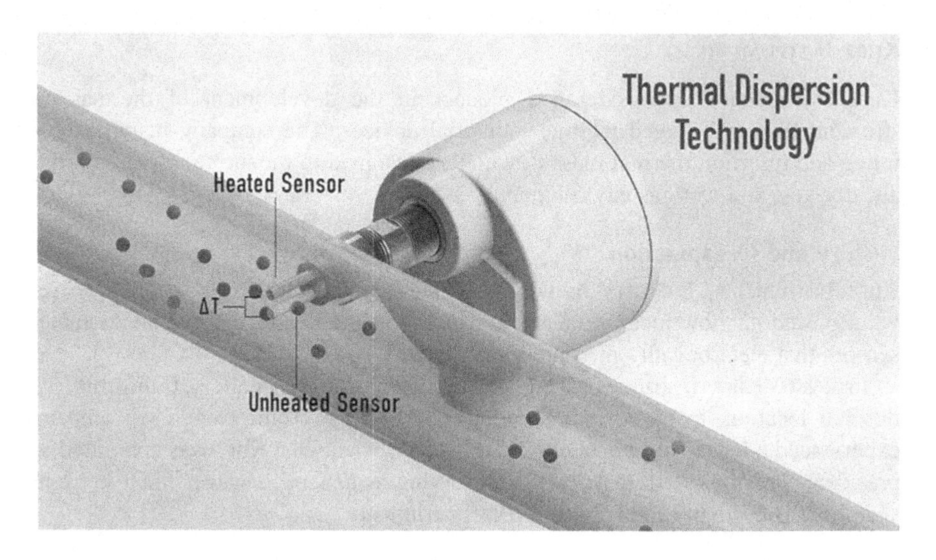

FIGURE 8.1 Thermal mass flowmeter technology. (Source: Graphic courtesy of Fluid Components Int'l.)

The second method is called a constant current method. Under this method, thermal flowmeters also have two sensors: one heated sensor and one that measures the temperature of the flowstream. Power to the heated sensor is kept constant. Mass flow is measured based on the difference between the temperature of the heated sensor and the temperature of the flowstream.

Both methods make use of the principle that higher velocity flows result in greater cooling. Both compute mass flow by measuring the effects of cooling on the flowstream.

Later Developments

Fluid Components, Kurz, and Sierra were the main companies manufacturing thermal flowmeters in the early 1980s. However, soon other companies came along. In 1988, Mark Eldridge founded Eldridge Products in Monterey, California, a company that focused on developing a variety of types of thermal sensors to measure the flow of various types of gases. Eldridge developed a multipoint thermal flowmeter that is deigned to be used in large cross-sectional areas where two or more measuring points are required. Examples are large intake ducts and air exhaust and flue stacks. In the early 1990s, Eldridge's multipoint flowmeters were developed for continuous emissions monitoring (CEM).

In 1994, Brad Lesko founded Fox Thermal Instruments in the Gilroy, California, also in the Silicon Valley area. Fox Thermal's flowmeters are used for a wide variety of applications, including boiler inlet, compressed air, submetering and custody transfer, digester gas, wastewater aeration, and flare headers. Fox distinguishes itself with research into thermal sensors and manufactures its own thermal sensors.

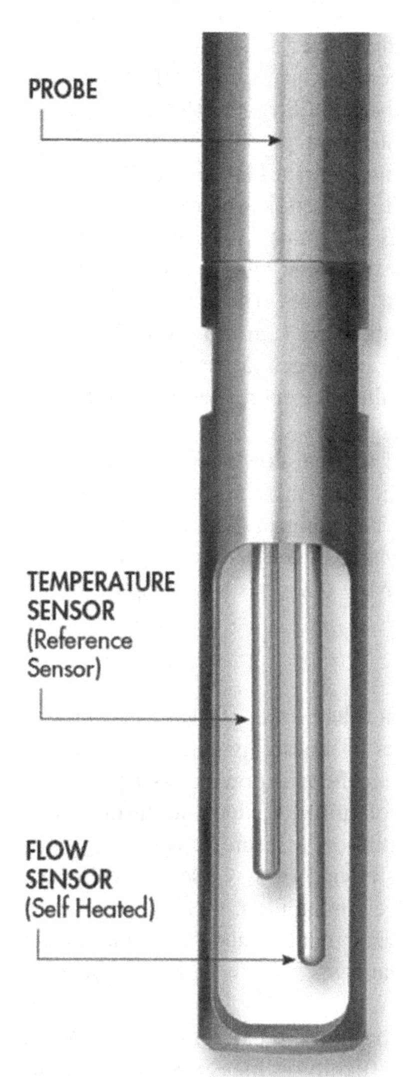

FIGURE 8.2 Example of dual temperature sensors to measure flowrate. (Source: Graphic courtesy of Sage Metering.)

In 2002, Bob Steinberg founded Sage Metering, after having worked for several other thermal flowmeter companies. Sage was founded in Monterey, California, where it still is located. Like the other thermal flowmeter companies described here, Sage focuses on gas flow measurement. However, Sage has developed a special focus on environmental applications, including measuring greenhouse gas emissions, flare gas measurement, carbon credits, and atmospheric storage tank vents.

Other Thermal Flowmeter Companies

Thermal flowmeters are often associated with Monterey, California, and the surrounding area because of the concentration of thermal flowmeter companies there.

However, there are a number of other thermal flowmeter companies in the United States and around the world. Thermal Instrument Company, located in Trevose, Pennsylvania, was among the first to introduce thermal flowmeter technology to the commercial marketplace in 1959. Thermal Instrument produces both inline and insertion meters, and it produces them to measure flow in both liquids and gases. Another US company to manufacture thermal flowmeters is Magnetrol.

In Europe, both Endress+Hauser and ABB are major suppliers of thermal flowmeters. Unlike most of the US companies described above, Endress+Hauser and ABB offer a wide range of instrumentation products. These include multiple types of flowmeters, pressure transmitters, temperature transmitters and sensors, and analytical instruments. Both these companies are much larger in terms of total company revenues and number of employees than the more thermal-centric US companies.

In Japan, Tokyo Keiso is a major supplier of thermal flowmeters. Tokyo Keiso introduced its first thermal flowmeter in 1977, its first thermal mass flowmeter in 1983, and its mini-thermal flowmeter in 1991. The company's thermal flowmeter line is focused on the measurement of gases. These flowmeters are used in waste incineration plants, in air conditioning, and in many other gas supply applications.

Advantages and Disadvantages

Thermal flowmeters have fast response time, and they excel at measuring flow at low flowrates. They also provide a direct means of measuring mass flow and can handle some difficult-to-measure flows. Insertion thermal flowmeters are used in CEM applications to help measure the amount of sulfur dioxide and nitrous oxide being released into the environment. Concentration measurements, along with flowrate measurements, are required. Thermal flowmeters have several other advantages. One is a relatively low purchase price. In addition, thermal flowmeters can measure the flow of some low-pressure gases that are not dense enough for Coriolis meters to measure. Both of these advantages give thermal flowmeters a unique niche in flow measurement.

One limitation of thermal flowmeters is that they are used almost entirely for gas flow measurement. Thermal flowmeters have difficulty in measuring liquid flows because of the slow response time involved in using the thermal principle on liquids. Some companies have released thermal flowmeters for liquid flow measurement, however.

The main disadvantage of thermal flowmeters is low to medium accuracy. Thermal flowmeters are not nearly as accurate as Coriolis meters. While some thermal flowmeters may achieve accuracy levels of one percent, other thermal flowmeters have accuracies in the three to five percent range. However, thermal suppliers are working to improve the accuracy of their flowmeters. Users who are considering thermal flowmeters need to balance their accuracy needs with their cost requirements. Expect wider use of thermal flowmeters as their accuracy levels increase (Table 8.1).

TABLE 8.1
Advantages and Disadvantages of Thermal Flowmeters

Advantages	Disadvantages
Medium cost	Low to medium accuracy
Ability to measure flow of low-pressure gases	Gas composition must typically be known
Well-suited for stack flow measurement	Almost entirely used for gas; very few liquid applications
Well-suited for emissions monitoring applications	Presence of moisture or droplets can lead to measurement inaccuracy
Insertion models can handle large pipe size measurement	Cannot be used to measure steam flow

Applications for Thermal Flowmeters

While thermal flowmeters are uniquely capable of supporting many gas flow applications, they have limited application for liquids, and they are not a good fit for steam flow measurement. Many of their applications are in process gas and other non-custody transfer applications. Thermal flowmeters do not have the necessary industry approvals for use with custody-transfer of natural gas in pipelines. This market is dominated by ultrasonic, turbine, and differential-pressure (DP) flowmeters, all of which possess custody-transfer approvals from the American Gas Association (AGA). Thermal flowmeters are unlikely to achieve this approval without technological breakthroughs that would increase measurement accuracy to the very high level required for custody transfer.

Today, thermal flowmeters compete with ultrasonic and DP flowmeters using averaging pitot tubes for flue gas monitoring. Flues typically are large pipes, stacks, ducts, or chimneys that dispose of gases created by a combustion process. Thermal flowmeters measure the flow of gases through flues.

Thermal flowmeters also compete with ultrasonic and DP flowmeters in flare gas monitoring. Flare systems are used to burn off waste gases from refineries, process plants, and power plants. Flares can be a single pipe or a complex network of pipes. Flares are subject to strict environmental regulations. Thermal flowmeters are used to measure the amount of gas flared (Figure 8.3).

GROWTH FACTORS FOR THE THERMAL FLOWMETER MARKET

This section discusses the growth factors underlying the thermal flowmeter market that specifically relate to positive growth in this market. Thermal flowmeter growth factors include:

- A new age of environmental awareness.
- Macroeconomic factors.
- CEM and greenhouse gas emissions measurement boost thermal flowmeter sales.

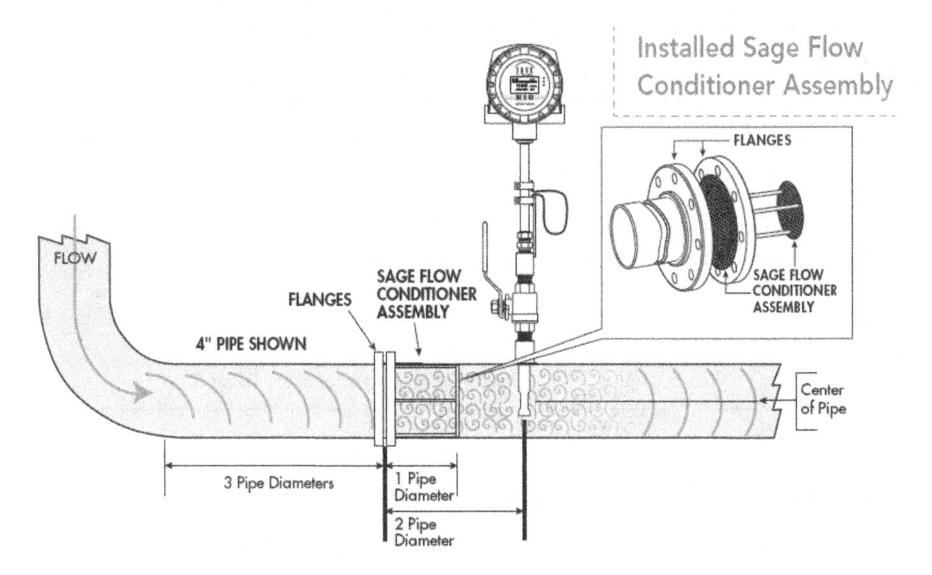

FIGURE 8.3 Typical Sage thermal mass flow meter installation with flow conditioner assembly. (Source: Courtesy of Sage Metering.)

- Low-flow gas measurement is a strength.
- High turndown ratios and pre-programming are valuable in many applications.
- Availability of in situ device verification.
- Ability to function in an integrated digital ecosystem.
- Flare gas measurement and submetering of gas flows.
- The water/wastewater industry.

A New Age of Environmental Awareness

Environmental protection is a major issue in all regions of the world and the overwhelming majority of countries. The current and future causes and effects of climate change have become centerpiece issues being addressed by governments and businesses alike, and the actions taken on the issues by these two institutions are being watched closely by both regulators and ordinary citizens alike.

At the very center of worldwide efforts are regulations governing air and water pollution, exhaust emissions, and waste disposal. Increasing concern with clean air and clean water – the two most basic components of our environment – will continue to fuel the growth of environmental regulations. These regulations are written and enforced in large part based on the measurement of gas and liquid flows. And it is the mutually agreed upon validity of the measurement outputs that permits producers of subject flows to prove their compliance with environmental regulations, and for regulators to determine who is and who is not in conformance with them.

Developing countries tend to be behind the most highly developed countries in the implementation of active environmental controls. This is because they often choose to put the emphasis on present economic development and defer the expense of environmental controls. Nonetheless, developing countries have and will continue to follow the lead of the more highly developed countries in environmental protection and control as their economies become more advanced and as public awareness of the importance of environmental responsibility increases. For this reason, countries in Latin America, Africa, and Asia are expected to increase their regulation of air and water quality, waste disposal, and other environmental matters. This will create increased demand for air and water quality measuring systems, many of which incorporate flow measurement and monitoring, including thermal flowmeters.

Macroeconomic Factors

Economic growth is driven by a variety of factors, but two are especially important. The populations of Latin America and Asia continue to expand at a high rate, and there is a continuing demand for a higher standard of living in the countries there – and important rises in income to do so. An expanding population means a demand for more chemical products, more power, more automobiles, more food to be processed and transported, more pulp and paper products, and more petroleum-based products. The demand for a higher standard of living generates a need for exactly the same result: more consumer goods being purchased by a population that has become used to a higher standard of living.

These forces of increased population and a higher standard of living have long been at work in the United States, Canada, and Western Europe, but are now most pronounced in Asia. Asia is composed of many developing countries whose current standard of living is significantly lower than that in North America or Western Europe. This means that these countries have farther to go before reaching a standard of living they view as satisfactory. In addition, many plants in Asia still have a significant installed base of electromechanical and pneumatic systems, especially in the power generation, metals, and paper industries. Thermal flowmeters are used in these industries. As companies in these industries face competition from abroad, they are converting to modern industrial systems. This will result in a growing need for thermal flowmeters, as well as other instrumentation.

An important growth factor underlying increases in the thermal flowmeter market is new plant construction. While not a lot of new manufacturing plants are being built in the United States, there is substantial new construction occurring in Latin America and Africa in addition to the high rates of construction in Asia, including China and India. These include chemical, power, pulp and paper, and food and beverage plants, as well as refineries. New manufacturing plants will in most cases install modern electronic instrumentation systems. In flow measurement, new technology flowmeters such as thermal will be more of a consideration because of their accuracy and their increased availability as a networked device, two important considerations in operating a plant facility cost-effectively.

CEM and Measuring Greenhouse Gas Emissions Boost Thermal Flowmeter Sales

While the need for Continuous Emissions Monitoring (CEM) is ongoing, the 21st century has brought new environmental awareness and requirements. Scientific thinking on the issue has evolved substantially in the past 20 years, and improved energy source alternatives have been developed. While global warming and the need to reduce carbon emissions were once viewed as merely plausible scientific theory, they are now almost universally accepted as scientific fact by the world community.

It was in the early 1990s that new environmental regulations began requiring companies to detect and reduce the emissions of sulfur dioxide (SO_2) and nitrous oxide (NO_x) into the air. SO_2 and NO_x were identified as two of the principal causes of acid rain. The Environmental Protection Agency (EPA) initiated federal programs in the United States to reduce pollution in the atmosphere. EPA regulations have resulted in the development of an entire industry around CEM.

In response to CEM requirements, thermal flowmeter companies developed multipoint thermal flowmeters. In many cases, CEM occurs in large stacks that emit pollution from industrial sources. Single point thermal flowmeters measure flow at a point, making it difficult to accurately compute flow in a large pipe or smokestack. Multipoint thermal flowmeters measure gas flow at multiple points and use these values to compute flow for the entire pipe, duct, or stack. Some multipoint flowmeters have as many as 16 measuring points.

The Kyoto Accord, brought into effect in 2005, is an international treaty designed to reduce greenhouse gas emissions, and has resulted in the creation of several mechanisms that require the measurement of greenhouse gases. These include Certified Emission Reductions (CER), which allow carbon emitters to gain mitigating credits for reduced carbon levels being emitted elsewhere. Another program is the Clean Development Mechanism (CDM), which allows countries to invest in sustainable development projects that reduce emissions in developing countries.

The Kyoto Protocol identified six of the main greenhouse gases:

- Carbon dioxide (CO_2)
- Nitrous oxide (N_2O)
- Methane (CH_4)
- Perfluorocarbons (PFCs)
- Hydrofluorocarbons (HFCs)
- Sulfur hexafluoride (SF_6)

Beginning in 2008, the Obama administration placed a renewed emphasis on identifying and curtailing the emission of greenhouse gases and set ambitious goals for making a move from fossil fuels to renewable energy. Europe initiated the European Union (EU) Emission Trading Scheme (ETS) to regulate the six gases listed above. So far, the main focus has been on carbon dioxide.

Since the Kyoto Accord there has been another pact with potentially even more far-reaching influence: the Paris Agreement of 2015, which entered into force in November 2016, and further codified the gases of most concern. The agreement included all of the nations of the world with just two exceptions and includes the stated intentions and goals of each signatory in addressing its own issues within climate change. The new age of environmental awareness that has spawned the Kyoto Accord, the Paris Agreement, and other greenhouse gas initiatives, has resulted in a rewriting of the rules on measuring greenhouse gas emissions.

The Trump administration attempted to reverse some of these trends, though with limited success. In a policy reversal, the United States rejoined the Paris Agreement in February 2021, after the election of Joe Biden. The Biden Administration is pushing aggressively for legislation on climate change and can be expected to increase the regulations governing greenhouse gas emissions. This presents an opportunity for thermal flowmeters and other types of flowmeters that can measure greenhouse gas emission. The Glasgow Climate Change Conference, held in November 2021, updated many provisions of the Paris Agreement, and resulted in renewed commitments from many countries to lower emissions and take action on climate change.

The need for greenhouse gas measurement is not limited to the United States; it is worldwide, and there is a need and a demand to measure greenhouse gases in applications that formerly may have gone unnoticed. Many of these applications present opportunities for thermal flowmeters, including the following:

- *Measurement and recovery of landfill gas:* Landfills produce carbon dioxide, methane, and a mixture of other gases. These gases are measured as they leave the landfills, extracted from different wellheads, and collected to a common header pipe. The collected gases are then disposed of or recovered as a fuel source.
- *Ethanol distillation and refining:* Ethanol production is a complex process involving both fermentation tanks and distillation tanks. Thermal flowmeters measure the flow of air and fuel going into the distillation tanks, and the CO_2 leaving the fermentation tanks.
- *Measuring emissions from steam generators, boilers, and process heaters:* Thermal flowmeters measure these emissions, especially NO_x and carbon monoxide.
- *Biomass gasification:* Organic industrial waste and food waste can be digested in aerobic conditions in reactor tanks and fermentation towers. The output from this process is called biogas. Biogas includes methane, carbon dioxide, and a mixture of other gases. Thermal flowmeters are used to measure gas flow at multiple points along the way, providing optimal production, control, and reporting.
- *Recovery of methane from coal mines:* Methane that is recovered from coal mines often is mixed with air, carbon dioxide, and nitrogen. It is important for accurate flow measurement to calibrate the flowmeter with the actual mixture of gases. Thermal flowmeters are used to measure the amount of extracted gases recovered from coal mines.
- *Monitoring of flue gas:* Flues are typically large pipes, stacks, ducts, or chimneys that dispose of gases created by a combustion process. Thermal flowmeters measure the flow of gases through flues. This is often required

by environmental regulations, but it can also be useful in determining the efficiency of the combustion.

- *Measurement and monitoring of flare gas flow:* Flare systems are used to burn off waste gases from refineries, process plants, and power plants. Flares can be a single pipe or a complex network of pipes. Flares are subject to strict environmental regulations. Thermal flowmeters are used to measure the amount of gas flared.

The applications noted here provide a mere sampling of the growing number of applications requiring measurement of greenhouse gases. Thermal flowmeters are uniquely suited to make these measurements because their insertion technology allows them to handle large pipe sizes and because they can accurately measure different mixtures of gases. The need for these measurements can be expected to grow substantially in the next five to ten years. As a result, so will the demand for thermal flowmeters.

Low-Flow Gas Measurement Is a Strength

Thermal mass flowmeters are primarily used in gas applications and are particularly useful in measuring low-flow conditions. Other flowmeter technology types, such as vortex and differential pressure, have difficulties in maintaining measurement accuracy when the flowrate falls below certain minimum levels, as they depend on either the creation of vortices in the flow or differences in upstream and downstream pressure in order to generate a flowrate based on fluid velocity.

Thermal mass flowmeters, on the other hand, directly measure mass flow versus volumetric flow based on heat transfer. Coriolis flowmeters are another technology that produces a direct mass flow measurement, but they are more expensive than thermal mass flowmeters and can be more difficult to install. Most manufacturers of thermal mass flowmeters have models that are highly sensitive to low flow rates. This inherent sensitivity is due to the basic operation of this technology, where one sensor measures the current fluid temperature as a reference and the second sensor is heated and has a constant temperature differential relative to the first sensor at "zero flow."

High Turndown Ratios and the Ability to Measure Multiple Gases Are a Plus

Thermal flowmeters typically have high turndown ratios of 100:1 or greater, making them highly suitable in applications that require consistent accuracy over a large range of flowrates. Examples of these applications include carbon dioxide (beverage production and chilling), compressed air (consumption, distribution), air or biogas measurement (wastewater plants), and natural gas (burner and boiler feed control). The production of oxygen and nitrogen also fits this category of need.

Many manufacturers produce thermal flowmeters that are pre-programmed to measure up to 20 individual gases, or gas mixtures. These gas types or mixtures can typically be selected in the field. Operators can adjust the flowmeter to the gas to be measured as needed or measure a different type of gas at another location without recalibrating the meter.

Flare Gas Measurement and Submetering of Gas Flows

The oil and gas industry is subject to much variability and is a highly visible example of the workings of supply and demand. The supply side – as represented by the exploration and production elements – is, essentially, a margin business where there is little value added to the product. It is not until the product (whether petroleum-based liquid or gas) reaches the refinery or gas processing plant that more complex physical and economic elements come into play, and then not until just before delivery to the consumer that there is any real value-add involved. Even at this point, the additives used in liquids (e.g., detergents, ethanol) or natural gas (e.g., mercaptan, thiophane) do not change the basic nature of the product, which is to be a combustible energy source.

Thermal flowmeters used within the oil and gas industry are thus indirectly subject to many of the same economic factors as the industry in general. Fortunately, thermal flowmeters are found most often in the midstream and downstream portions of the value chain, where demand for petroleum-based products has remained relatively constant even as available supplies of petroleum fuel-based supplies increased. This condition is likely to remain in force as long as thermal flowmeters are not approved for custody transfer and other flow technologies – for their own other unique characteristics – remain preferred choices. Ultrasonic and pitot tube–based differential pressure technologies are direct competitors to thermal flowmeters here.

Applications in flare gas measurement are a good example of where thermal flowmeters do well in oil and gas. Gas flaring had been commonly used at one time as a method of disposing of excess combustible gases or to relieve the gas pressure within a system. Before global warming became an issue, this method was a widely accepted and economical practice. There were no concerns regarding environmental impacts and no real incentives to maximize the energy content of the gas being flared.

All the above reasons to not be more mindful of the consequences of gas flaring have changed since the early 1990s. In today's world, there are numerous incentives to avoid flaring gas, and some rewards for not doing so. Natural gas is a valuable resource, and businesses now seek to maximize its value by recycling excess quantities rather than destroying them. In addition, there is the pervasive regulatory environment mentioned above that encourages the reduction in emissions of burned gases through sets of both incentives and penalties. Businesses now see the value in the proposition of reducing their greenhouse gas emissions while acquiring a new source of onsite energy.

The cycle of measuring, recording, and tracking flare and stack gas emissions is likely to intensify over time, and thermal flowmeters – especially the multipoint insertion versions – should be a prime beneficiary of this trend.

Another area with good growth prospects is submetering of gas flows. Any large user of natural gas for heating is a potential customer, and the list includes settings such as commercial and industrial buildings, campus environments such as hospitals or schools, and multi-unit residential complexes. The premise of the application is that energy is a genuine cost center for users of multiple application types, and its accurate measurement should be reliable and fully accountable.

When facility managers use a thermal mass flowmeter for submetering, they provide themselves a verification of the utility company's bill for actual natural gas consumption. In addition, this same submetering application can be extended within the facility or campus to make the organization's departments, individual campus buildings, or even tenants more aware of and responsible for their use of energy.

The Water and Wastewater Industry

The water and wastewater industry is another key segment driving thermal flowmeter success. The worldwide demand for clean potable water continues to increase due to population increases, the rise in expectations for healthful water sources by these populations as a whole, and the general purposes of accelerating industrial and commercial growth. Wastewater needs are the natural consequences of water usage, and these needs include the requirement of proper disposal of used water or, increasingly, its re-use. Water is less of a commodity than ever. Water consumption and treatment are becoming more specialized.

Thermal flowmeters have established themselves in this industry and have been a replacement choice for traditional technologies such as differential pressure as they, for instance, do not introduce pressure drop into the process flow and require less maintenance.

Wastewater treatment applications abound. One of the more common is the measurement of air/oxygen gas used to promote the secondary treatment of sludge. This step in the treatment process is important, as it is used to decompose the sludge by way of introducing "activated" sludge (i.e., sludge containing specific aerobic bacteria) that speeds the decomposition and thus increases process throughput. Careful air/oxygen measurement ensures that this step is conducted within ideal parameters, and that no energy is unnecessarily wasted in pumping more air or oxygen than the process requires.

Further downstream within a treatment plant, thermal flowmeters can be found in distribution pipes and aeration basins. And, on heading toward the output side, the decomposed sludge is exposed to anaerobic treatment using other specific bacteria chosen for this purpose. The result of this step is the production of water and a mixture of gases, primarily carbon dioxide and methane. The latter, also called digester gas or biogas, is a growing source of a type of renewable energy. It has come into use to power on-site plant operations and is also available as a commercial product.

FACTORS LIMITING THE GROWTH OF THE THERMAL FLOWMETER MARKET

This section discusses factors that limit the growth of the thermal flowmeter market, including:

- Limited application for liquids and steam
- Sensitivity to changing gas compositions/properties
- Lack of approvals for custody transfer

Limited Application for Liquids and Steam

While thermal flowmeters are uniquely capable of supporting many gas flow applications, they have limited application for liquids, and they are not a good fit for steam flow measurement. This explains why thermal technologies account for such a small share of the overall flowmeter market, as gas flow accounts for only about 25 percent of total flow measurement applications. Many of these applications are in process gas and other non-custody transfer applications.

Sensitivity to Changing Gas Compositions/Properties

Thermal flowmeters are calibrated for the gas or gas mixture to be measured when deployed in the field. Every gas/gas mixture has its own unique properties that will directly correlate to its dissipation or absorption of thermal energy, all of which affect heat transfer characteristics. Thermal flowmeters are thus calibrated for specific gases or gas mixtures. And in the case of gas mixtures, the actual gas mix is passed through the flowmeter multiple times during the course of its calibration to establish the relationship between mass flow and the signal for the gas and sensor being calibrated. This process adds some cost to the eventual delivered product.

Another difficulty that users experience with thermal flowmeters is when they encounter conditions where there is moisture in the gas being measured. The presence of water droplets on the device's resistance temperature detectors (RTDs) prevents an accurate measurement from taking place, as the instrument's operational presets do not take water into account, only dry gas.

Even given the above, non-condensing water vapor may be acceptable in the process, though this is not a recommended condition. This, again, is because the heat loss to a liquid such as water droplets is so much greater than the heat loss to a dry gas that the flowmeter's flow signal will typically rise to a higher value. This introduces a flow measurement error that remains until the heated RTD is dry once again.

Lack of Approvals for Custody Transfer

Thermal flowmeters do not have the necessary industry approvals for use with custody-transfer of natural gas in pipelines. This market is dominated by ultrasonic, turbine, and differential pressure flowmeters, all of which possess custody-transfer approvals from the American Gas Association and other international agencies. Thermal flowmeters are unlikely to achieve this approval without technological breakthroughs that increase measurement accuracy to the very high level required for custody transfer.

In the meantime, Coriolis flowmeters have become a more economical option in the lower line sizes but charge a price penalty in the largest size lines that make them uneconomical for most purchasers. In the foreseeable future, thermal mass flowmeters do not seem to have a place in this gas market segment that will continue to be dominated by ultrasonic, turbine, and differential pressure technologies.

FRONTIERS OF RESEARCH

Improved Accuracy

Thermal flowmeters have never been able to achieve the same accuracy levels as ultrasonic, Coriolis, or even magnetic flowmeters. One reason is that the vast majority of thermal flowmeters measure gas, and gas is inherently more difficult to measure than liquids. Expect thermal suppliers to continue to work on improving the accuracy of their flowmeters.

Measuring Greenhouse Gases

Thermal flowmeters are well suited to a variety of environmental applications, including measuring biogas, biomass, boiler emissions, fuel cells, and many other applications. Thermal flowmeters are also suited for measuring flare gas and stack gases. These applications are a key to the expansion of the thermal flowmeter market. Expect suppliers to continue to focus on them as they seek to expand market share.

9 Application Advances in New-Technology Flowmeters

OVERVIEW

In previous chapters we discussed the frontiers of research for particular flowmeter technologies. However, many frontiers of research are not unique to a single technology; instead, they are common to a number of flowmeters. For example, the efforts to improve accuracy in measurement are ongoing in all the new-technology flowmeters, and in many conventional flowmeters as well. Accuracy is a critical value for custody transfer measurement, especially for high-value fluids. Other important frontiers of research for multiple types of flowmeters include reliability, cost, redundancy, materials of construction, sanitary applications, and communication protocols. End-users may well select their type of flowmeter based on how the flowmeters perform in one or more of these areas. Suppliers also compete with each other to outperform the competition in accuracy, reliability, cost, and other important features. It is no wonder, then, that these areas are frontiers of research for many types of flowmeters.

CUSTODY TRANSFER

The ability to do custody transfer measurement is one of the most highly prized features of a number of flowmeter types. Custody transfer measurement occurs when the ownership of a fluid is transferred from one party to another. When high value fluids such as crude oil or natural gas are bought or sold, accurate measurement is vital. Payment is generally made based on the amount of fluid transferred, and even a small measurement error can quickly be magnified over time. Emerie Dupuis of Daniel Measurement and Control illustrates this as follows:

> For example, Pump Station 2 on the Alaska Pipeline is designed to pump 60,000 gallons per minute (227 cubic meters per minute) of oil. A small error of 0.1% equates to an error of 2,057 barrels of oil a day. At a spot price of $105 a barrel, that 0.1% error would cost $216,000 a day. Over a year, the 0.1% error would amount to a difference of $78.8 million. Note that the error could either be on the high side, benefiting the seller; or on the low side, to the buyer's benefit. (*Oil and Gas Custody Transfer,* by Emerie Dupuis in *Petroleum Africa,* May 2014).

Custody transfer occurs often in the oil and gas industry. The owner could, for example, be a pipeline company, an oil or gas production company, or a utility company. In a flow

DOI: 10.1201/9781003130017-9

measurement custody transfer situation, what typically happens is that one or two custody-transfer flowmeters measure the volume or mass of fluid before the transfer is made, and then another set of flowmeters measures the flow after the transfer. Custody transfer is unique among flowmeter applications in that money changes hands and accuracy requirements are higher than they are for most other applications.

The ability to perform custody transfer measurements depends on several criteria. One is the approval by governing bodies such as the American Gas Association (AGA) or the American Petroleum Institute (API). A second criterion has to do with meter performance. Flowmeters used for custody transfer measurement must meet certain performance criteria, including accuracy.

Standards

The AGA began studying custody transfer for natural gas applications in the late 1920s. Its first report was issued in 1930 and was called AGA-1. The subject of AGA-1 was the use of differential-pressure flowmeters with orifice plates for custody-transfer applications. This report was followed by AGA-3, which was first issued in 1955 and reissued in 1992. The AGA issued a report in 1981 on the use of turbine flowmeters for custody-transfer applications. This report was called AGA-7 and it applied to gas applications. The report was reissued in 2006.

The AGA published a report on the use of Coriolis flowmeters for custody-transfer applications, AGA-11, in 2003. In the mid-1990s, the move to standardize the use of ultrasonic flowmeters for custody transfer began in Europe. At that time, Groupe Européen de Recherches Gazières (GERG) published Technical Monograph 8. This report laid out the criteria for using ultrasonic flowmeters for custody-transfer applications. Following this, the AGA published AGA-9 in 1998. This report also specified custody transfer applications for ultrasonic flowmeters. Though it took some time for this standard to be widely accepted, it greatly increased the use of ultrasonic flowmeters for custody transfer, especially for natural gas pipeline applications.

While the AGA and the API work together on many standards projects, the AGA is more focused on industrial and natural gas, while the API deals more with petroleum liquids. In fact, the API has issued its own reports on the use of flowmeters involving custody transfer of liquids. These include API MPMS 5.2 on positive-displacement meters, API MPMS 5.3 on turbine meters, API MPMS 5.6 on Coriolis flowmeters and API MPMS 5.11 on ultrasonic flowmeters. Other API reports focus on the use of magnetic, thermal dispersion, and variable area flowmeters.

While differential pressure, turbine, Coriolis, and ultrasonic flowmeters all have standards approval from the AGA and API, vortex flowmeters have lagged behind in terms of getting approvals. The API took the first step toward remedying this situation when in January 2007 an API committee approved a draft standard for the use of vortex flowmeters for custody transfer of liquid and gas. This standard was updated in 2010. In 2013, the standard was looked at again. Finally, in 2017, it was approved as:

> Manual of Petroleum Measurement Standards (MPMS) Chapter 14 – Natural Gas Fluid Measurement Section 12 – Measurement of Gas by Vortex Meters.

While this is a major hurdle cleared for vortex meters, it is taking time for this standard to have an impact on vortex flowmeter sales. This also happened after ultrasonic meters were approved for custody transfer. As time goes on, and people become more familiar with API 14.12, they will take advantage of the use of vortex meters for custody transfer of gas.

Performance

While flowmeters need standards approvals to qualify for custody transfer applications, performance is another important criterion for flowmeters to be used for this purpose. In particular, flowmeters need to be sufficiently accurate to meet custody transfer requirements. Whether a flowmeter can perform at a level sufficiently high to reliably perform custody transfer measurement depends on its technology, its construction, its calibration, its installation, the fluid being measured, and other factors.

The accuracy of a flowmeter is its ability to produce a flowrate, volumetric flow, or mass flow that corresponds to the actual flowrate, volumetric flow, or mass flow of the fluid flowing through the pipe whose flow is being measured. The same definition applies to open channel flow, except that instead of pipe flow being measured, flow through an open channel is being measured. An open channel could be a partially filled pipe – closed pipe flow is always flow under pressure. Repeatability refers to the ability of the flowmeter to produce approximately the same reading consistently when the same variables are present. Typical variables include temperature, pressure, flow profile, and other relevant conditions.

A flowmeter can be highly repeatable but nonetheless inaccurate if it repeatedly produces the same inaccurate reading under the same conditions time after time. On the other hand, an accurate flowmeter is also repeatable. If a flowmeter is accurate, it will produce the same accurate reading under the same conditions time after time. If the reading varies beyond a certain narrow boundary, it will no longer be accurate or repeatable.

There are a number of criteria for evaluating the performance of a flowmeter. These include accuracy, repeatability, fluid types measured, longevity, communication protocols, materials of construction, and many other factors. There are many ways to determine flowrate, and some are more accurate than others. Sticking your finger in a flowing stream (including air) is one way, though this is not a reliable method. Each of the flowmeter types considered in this series of books on flow measurement uses a specific technology.

Probably the single most important factor affecting the performance of a flowmeter is its technology. Coriolis and ultrasonic flowmeters perform at a higher level than any other new-technology flowmeters. Coriolis meters use the momentum of the fluid and how this impacts one or more oscillating tubes to determine mass flow. Ultrasonic meters base their reading of flowrate on difference in transit time between ultrasonic signals traveling upstream and those traveling downstream. Differential pressure meters place a constriction in the flowstream and measure the difference between upstream and downstream pressure to compute flowrate.

The technology used by each type of flowmeter has a determining impact on its performance. Coriolis flowmeters achieve the highest accuracy of any flowmeter type because computing mass flow based on fluid momentum yields a highly accurate mass flow measurement. Likewise, transit time ultrasonic meters produce a highly accurate, repeatable, and reliable flowrate measurement. Variable area meters, by contrast, rely on the height of fluid as shown by a scale, somewhat like a ruler, that is applied to the side of the meter. This method is inherently less accurate than the methods used by other flowmeter types because it involves a manual reading comparing the fluid level to the scale on the side of the meter. Most variable area meters do not have an output, although some are now being made with an electronic output. Even with an output, variable area meters are less accurate and less repeatable than other flowmeter types.

Coriolis Flowmeters

Many, though not all, Coriolis flowmeters perform at a sufficiently high level to meet the criteria for custody transfer measurement. Coriolis flowmeters used for custody transfer have accuracies as high as 0.1 percent or even 0.05 percent. An exception is Emerson's R-Series, which is designed to be a low-cost Coriolis meter. The accuracy of these meters is in the 0.5 percent range. Other suppliers have brought out low-cost Coriolis meters. This shows that it is not just the technology but also the performance of a flowmeter that determines whether it can be used for custody transfer applications.

While Coriolis flowmeters can perform at a high level for custody transfer applications, not all Coriolis meters are equally suited for custody transfer. Whether a Coriolis meter is suitable for custody transfer depends on the type of fluid being measured, and on whether it is dual or single bent tube, or single or dual straight tube. Other factors include proper calibration, proper installation, and line size. With the price of Coriolis meters ranging from $5,000 to $75,000, depending on tube type, line size, and materials of construction, not all Coriolis meters perform at custody transfer levels. However, there are many other types of measurement, including allocation metering, in-plant measurement, hydraulic fracking, and chemical blending, that do not involve custody transfer and hence do not require custody transfer level of performance.

Ultrasonic Flowmeters

Only multipath (three paths or more) ultrasonic flowmeters qualify for custody transfer applications. This is true for ultrasonic flowmeters measuring gases and liquids. Ultrasonic flowmeters used for custody transfer have accuracies in the range of 0.15 percent or even 0.1 percent. This applies to inline (spoolpiece) flowmeters, not to clamp-on or insertion meters. It also does not apply to Doppler meters. Ultrasonic flowmeters are widely used for custody transfer of natural gas, especially for measurement of natural gas in large pipelines (e.g., 20–42 inch diameters). They are also used for custody transfer of petroleum liquids.

The most accurate ultrasonic flowmeters are inline (spoolpiece) transit time meters with three or more paths. Ultrasonic meters can achieve accuracy levels that are very close to those of Coriolis meters. For custody transfer purposes, ultrasonic meters must have at least three or more paths; single and dual path ultrasonic meters cannot reliably perform custody transfer measurements. Ultrasonic meters with three or more paths are known as multipath meters. Multipath meters commonly have four, five, or six paths, depending on the manufacturer, but others are built with 8, 12, and even 18 paths.

Vortex Flowmeters

Vortex flowmeters have been approved for use in the custody transfer of gas. The most accurate vortex flowmeters achieve accuracy in the 0.5 percent range, though many are less accurate than that. One issue that vortex meters have is in measuring low flowrates. The flowstream must be moving at a minimum rate in order to generate the vortices required to measure flow. Another issue for vortex meters has been vibration. If vibration occurs in the line, it can create disturbances in the flowstream that can interfere with accurate measurement. A number of supplies have addressed this issue with advanced software and advanced signal processing.

Magnetic Flowmeters

Magnetic flowmeters cannot measure gas or hydrocarbons, but they have received some custody transfer approvals from the American Water Works Association (AWWA) for downstream delivery of water. Here they are competing with positive displacement and turbine meters. Magnetic flowmeters have not achieved the same accuracy levels as Coriolis and ultrasonic, and for many magmeters, accuracy is in the 0.5 percent range.

Thermal Flowmeters

In terms of the other new-technology flowmeters, thermal meters still top out around 1.0 percent in terms of accuracy and have not yet been approved for custody transfer measurement. Instead, they rely on non-custody transfer measurements. Examples include measurement of flare gas and stack gas emission, continuous emissions monitoring (CEM), and measurement of greenhouse gas emissions. In flare and stack gas emissions, thermal flowmeters compete with multivariable differential pressure and ultrasonic flowmeters.

ADVANCES IN FLOWMETER TECHNOLOGY

One of the ways that advances are made in flowmeter technology is when suppliers seek to enhance the accuracy, repeatability, and reliability of the flowmeter types they manufacture. This can be done by using stronger materials of construction like titanium, by using enhanced signal processing, by introducing higher levels of computing power into the transmitter, by creating more sensitive sensors, and by adding new communication protocols to the flowmeters. These are just some of the many ways that advances in flowmeter technology are made. The way in which flowmeters are made more accurate, repeatable, and reliable depends on the technology they use. Some of these methods are discussed in the following section.

Coriolis Flowmeters

Coriolis flowmeters are generally viewed as being the most accurate meter; they can achieve accuracies of 0.1 percent, or in some cases, 0.05 percent. This applies when the fluid being measured is liquid and not gas. For gas, Coriolis flowmeters typically achieve accuracies in the range of 0.5 percent, although 0.25 percent is possible. Dual bent tube Coriolis generally perform better than single bent tube meters and bent tube meters generally perform at a higher level than straight tube meters. However, straight tube meters are often preferred for sanitary applications, since fluid can build up around the curves of bent tube meter. Straight tube meters are also easier to drain.

One important factor in a Coriolis flowmeter is the shape of the tube. Coriolis designers have experimented with different tube shapes over the years. Certain tube shapes were selected to minimize pressure drop, maximize sensitivity, or achieve high frequency or compact size. In some cases, tube shapes were designed to avoid patent infringement. Two tubes work better than one because they create a balanced system, somewhat like a tuning fork. In a dual tube arrangement, one end of the pair of tubes is attached to a flow splitter, which splits the flow between the two tubes. A temperature sensor is placed along the tube to monitor the temperature of the tube. This is important so that a small compensation can be applied due to temperature changes in the material.

Straight tube meters were designed to avoid some problems with bent tube meters. Bent tube meters are hard to clean, and fluid can build up along the curvatures in the tubes. Endress+Hauser introduced the first commercially successful single tube straight tube meter in 1987. In the early 1990s, Schlumberger had a dual tube straight tube meter that it attempted to bring onto the market. This effort was not successful, and the meter was withdrawn. In 1994, KROHNE brought the first dual tube straight tube meter onto the market. Today the dual bent tube design is still the most dominant design. However, single and dual straight tubes have found a place in the market, especially for sanitary applications. They are most widely used in the food and beverage, pharmaceutical, and chemical industries.

The emergence of large line size (from 8 to 16 inches) Coriolis meters was discussed in Chapter 4. If the 16-inch barrier is to be broken, it will have to be with a material of extremely light weight combined with greatly enhanced sensor detection and signal processing. It could also be that someone will find a way to make a Coriolis measurement on the model of mass flow controllers, where a portion of the fluid is diverted from the flowstream and measured, and then fed back into the flowstream. It will be interesting to see who, if anyone, is the first to break this line size barrier.

Ultrasonic Flowmeters

Like Coriolis meters, ultrasonic flowmeters come in a wide range of configurations. There are three mounting types: inline, clamp-on, and insertion. There are also two main technologies: transit time and Doppler, as well as a hybrid that combines the two. Another important variation in ultrasonic flowmeters is the number of paths, which can vary from 1 to 18. Because ultrasonic flowmeter measurements depend on the speed of sound, different meters are required for measuring gases than for measuring liquids.

There are many applications for ultrasonic flowmeters that do not involve custody transfer. These include measuring flare gas, water and gas utility measurement, leak detection, in-plant measurement, batch dosing of liquids, and check metering. Doppler ultrasonic meters can measure the flow of dirty liquids, and here they compete with magnetic flowmeters. Clamp-on meters are widely used for check metering purposes.

One area of ongoing research is the number of paths in ultrasonic flowmeters. Single and dual path ultrasonic meters are not able to do custody transfer measurements and are generally used for applications requiring lower accuracy. Traditionally, Daniel has had four-path, Elster has had five-path, and FMC Technologies (now TechnipFMC) has had six-path meters. These meters are designed for gas flow measurement. KROHNE has a three-path ultrasonic meter and a five path ultrasonic meter, both designed for custody transfer of crude oil, refined fuels, and other applications. Caldon (now part of Sensia, a joint venture between Schlumberger and Rockwell Automation) markets the eight-path LEFM for petroleum liquids, although it was originally intended for nuclear applications.

The following section discusses the history of Daniel, Elster, and TechnipFMC.

Daniel Measurement and Control

Daniel was founded in 1946 by Paul Daniel, inventor of the Daniel Senior Orifice Fittings. Paul Daniel, born in 1894 in Houston and educated in a one-room schoolhouse, started his career in the oil industry in Southern California in 1915 at a refinery operated by Standard Oil Company. Later, when out of work during the Great Depression, Daniel designed the now-famous orifice fitting to enable plates to be changed within the oil pipeline system without interrupting the flow of the oil or allowing significant leakage. A few years later he formed Daniel Orifice Fitting Company. By the time Daniel left active management in the mid-1960s, the company had more than 500 employees and the company's product line included piston-controlled check valves, orifice flanges, and Simplex plate holders. In 1966, the board of directors changed the company's name to Daniel Industries, Inc.

Emerson Electric Company, a global technology and engineering company, acquired Daniel Industries for $460 million in 1999 to bolster Emerson's presence in the oil and gas industry. The Daniel brand within Emerson's Automation Solutions business segment aimed to deliver advanced insight and reliable measurement for the most challenging fiscal measurement applications. In addition to liquid turbine meters, the Daniel brand offered ultrasonic flowmeters and primary elements for both gas and liquid applications.

On July 12, 2021 Emerson announced a definitive agreement to sell its Daniel Measurement and Control Business to Turnspire Capital Partners by the end of Emerson's fiscal year. The sale, announced as final on August 31, 2021, includes all of Daniel's brand rights, facilities, intellectual property, and personnel, but does not include Daniel's ultrasonic flowmeters, which remain with Emerson.

Turnspire invests in high-quality businesses that have reached "strategic, financial or operational inflection points." The firm aims to make its companies best in class in their individual niches through operational improvements rather than financial leverage. Its stated mission is to "provide creative solutions to complex problems."

Honeywell Elster

For many years, Elster has been a major supplier of ultrasonic flowmeters. Elster can trace its roots to three events. The first was the founding of the American Meter Company (AMCO) in 1836, the second was the formation of Elster Meters in Berlin in 1848, and the third was the founding of Instromet in the Netherlands in 1965. Elster was originally a maker of gas lamps and began production of large station gas meters in 1853 in order to grow the company.

In 1985, Ruhrgas Industries acquired Elster Gas Metering and continued to acquire a larger portfolio of metering companies, including AMCO, in 1988, and Instromet, in 2001. In 2005, funds advised by CVC Capital Partners acquired Ruhrgas and renamed it Elster Group. In September 2010, Elster became a publicly traded company following its initial public offering. In 2012, Elster Group was acquired by Mintford AG, a wholly owned subsidiary of Melrose PLC. And finally, in 2015, Elster Group was acquired by Honeywell.

Elster is a leading supplier of ultrasonic gas flowmeters worldwide. The Honeywell Elster portfolio of industrial process control ultrasonic flowmeters has standardized on transit time technology. They are engineered for regulating industrial gas processes, and for the flow measurement of a wide range of combustible and non-combustible gases.

The company's most recent ultrasonic product is its Q.Sonic^{plus} gas flowmeter. A sophisticated six-path design, and the product of an "enhanced" Elster-Instromet patent, it is delivered with more functionality and with greater processing power that lowers measurement uncertainty. It is available in line sizes ranging from 3 to 56 inches and is suitable for custody transfer as well as gas exploration, transmission, and distribution applications.

Elster's Q.Sonic meter is available in three-path, four-path, and five-path configurations. The most advanced five-path models represent a sophisticated combination of transducers, digital electronics, and a unique and patented path configuration. The Q.Sonic five-path provides high accuracy (optionally, at better than ±0.5 percent of flowrate) and can analyze swirl and asymmetrical flow profiles. The Q. Sonic five-path has been used for custody transfer applications by numerous natural gas pipeline operators in the United States.

The TwinSonic^{plus} is a patented six-path design suitable for gas custody transfer applications. Using reflective paths and sophisticated diagnostics, the flowmeter can identify fouling or liquids inside the pipe. Fiscal measurements are fully compliant with AGA-9 and explosion approved according to ATEX, IECEx, FM, and CSA. Continuous verification of the measurement is achieved using additional 100 percent redundant electronics and dual path configuration built into the meter body.

FMC Technologies/TechnipFMC

FMC Technologies traces its roots to 1884 when inventor John Bean developed a new type of spray pump for his orchard and started the Bean Spray Pump Company. Mergers in the late 1920s with makers of vegetable processing equipment and cannery machinery resulted in a name change to Food Machinery Corporation (FMC). In 1961, the company changed its name to FMC Corporation.

In 1995, FMC Technologies acquired Smith Meter, to enhance a portfolio of measurement equipment that already included Kongsberg Metering and F.A. Sening. FMC has continued to develop the Smith Meter turbine and positive displacement flowmeters and manufactures all Smith Meter brand products in Erie. In 1999, FMC Technologies acquired the flow measurement assets of Perry Equipment Corp. (PECO) of Texas. The company's orifice gas measurement products complemented the Kongsberg ultrasonic gas meter and strengthened the Smith measurement systems business. FMC Technologies then reorganized the combined products and services as its Measurement Solutions division.

FMC Technology's ultrasonic flowmeter was called the MPU. The origins of the six-path design that later became the MPU 1200 go back to the early 1990s. This design was developed by Fluenta/CMR. Fluenta was created by CMR in 1985 as an industrial outlet for its inventions. Kongsberg Offshore was a part owner of Fluenta. In 1993, Fluenta/CMR delivered a prototype of the six-path design ultrasonic flowmeter to Statoil. In 1997, FMC acquired the ultrasonic technology from Fluenta, and delivered the first ultrasonic flowmeter to Statoil, then named the FMU 700. Soon after, FMC introduced the MPU 1200 version, which incorporated the second-generation electronics. A third generation of transmitter electronics for the meter, with enhanced signal processing, followed in 2002. This third-generation technology was developed to accommodate both liquid and gas ultrasonic meters and coincided with the introduction of FMC's first liquid ultrasonic flowmeters in 2002.

The Value of Extra Paths

According to Steve Caldwell, who became co-owner of Colorado Engineering Experiment Station, Inc. (CEESI) in 1986, adding more paths to an ultrasonic flowmeter does not in itself result in higher accuracy. CEESI is a flow calibration facility that is world renowned for its testing and calibration of flowmeters. On the other hand, there is no doubt that multipath ultrasonic meters, especially those with four or more paths, are more accurate than single or dual path ultrasonic meters. In some flowmeters with multiple paths, not all the paths are used to compute flowrate. Instead, the additional number of paths enables the flowmeter to take into account flow profile and turbulence and provides additional diagnostic capabilities. This includes the capability of detecting build-up on the pipe wall, which can affect accuracy in measurement.

In addition to the number of paths, there are many other areas of research where suppliers have made advances. These include transducer design, the positioning of the transducer in the pipe wall, signal processing of the ultrasonic signals, material of construction of the meter body, and other factors. In clamp-on meters, suppliers have introduced technology to measure the thickness of the pipe wall in order to take into account the attenuation of the ultrasonic signal as it passes through the pipe. These are just some of the many areas in which technological advances have been made in ultrasonic flowmeters since their introduction in 1963.

Vortex Flowmeters

Vortex flowmeter technology is described in Chapter 4. Vortex meters rely on a cylindrical object called a bluff body that is inserted into the flowstream. This bluff body generates vortices. The frequency of the vortices generated is proportional to flowrate. Vortex frequency is determined by means of sensors downstream from the bluff body. This value is sent to the transmitter, which uses it to calculate flowrate.

Suppliers have made many advances in vortex flowmeter design over the past ten years. Vortex meters have become more stable, more accurate, and more reliable as a result. For example, Emerson uses a non-clog design for its vortex meters by making the meter body using welded, gasket-free construction that has no ports or crevices that can clog. Yokogawa, Emerson, and other companies have addressed the problem of vibration by using advanced signal processing and by mass balancing of the sensor system. Built-in diagnostics in some meters enable the field verification of the meter electronics and the sensor without shutting down the process.

As discussed in Chapter 7, Jim Storer invented the multivariable vortex flowmeter in 1996 and founded VorTek Instruments for the purpose of manufacturing it. Since that time, many of the main suppliers of vortex meters have introduced their own multivariable vortex meters. Multivariable vortex meters contain a temperature sensor and a pressure sensor that enable the calculation of mass flow for both steam and liquids. Many of these multivariable flowmeters can measure both saturated and superheated steam. Vortex flowmeters are ideal for steam flow measurement because they can tolerate the high temperatures and high pressures that typically characterize steam flow. Their main competitors for measuring steam flow are differential pressure flowmeters. Superheated steam represents only about 4 percent of vortex measurements, while saturated steam represents just over 30 percent. The balance of the fluid measured by vortex meters is a combination of gas and petroleum and non-petroleum liquids.

Another area where suppliers have introduced variations is in the bluff body design. Designs typically are rectangular, square, trapezoidal, or t-shaped. If a bluff body is too thin, it may not create sufficient vortices. On the other hand, too wide a bluff body may create a pressure drop. The length of the bluff body needs to be in correct proportion to the bluff body's width. Vortex meters already have a problem measuring low flowrates, and this needs to be taken into account in bluff body design.

Several different sensing technologies are used to detect vortices. Over 70 percent of vortex sensors are piezoelectric, while a significant amount of the rest are capacitance sensors. A small percentage of vortex suppliers use ultrasonic sensors. Piezoelectric and capacitance sensors detect the pressure oscillations around the bluff body. Some vortex sensors are located inside the bluff body, while other are placed outside it. Sensors placed outside the bluff body are wetted and are enclosed in hardened cases to avoid the effects of corrosion or erosion from the fluid.

Body material is another area where suppliers have made advances. The meter bodies of over 80 percent of vortex meters are from 316 steel. This metal provides good corrosion resistance. Other materials used for meter bodies are Hastelloy, carbon steel, plastic, and Inconel. Hastelloy is especially resistant to acids, while Inconel resists corrosion and does especially well in high temperatures or high pressures.

Magnetic Flowmeters

Magnetic flowmeters work by using wire coils that are mounted onto or outside of the meter body. When current is applied to these coils, an electromagnetic field is created in the area where the fluid passes. As the fluid passes through the electromagnetic field, a voltage is created. Two electrodes on either side of the pipe sense this voltage, and pass a signal to the transmitter. The transmitter uses the signals from the electrodes to generate a flowrate. This flowrate is then passed to a controller, a distributed control system, or a recorder.

Magnetic flowmeters are popular because they are obstructionless and because they provide a cost-effective way to measure chemicals, slurries, and dirty liquids. They also provide a highly accurate measurement. Magmeters can measure hard-to-measure liquids that few other flowmeters can accurately measure. The main competitor to magmeters for these types of liquids is Doppler flowmeters. Positive displacement flowmeters can also measure dirty liquids, but they are mainly used for petroleum-based liquids. Magnetic flowmeters cannot measure hydrocarbons because they can only measure conductive liquids, and hydrocarbons are not conductive.

For many years, magnetic flowmeters have been the top revenue producer for worldwide sales. While this is still true in 2022, Coriolis meters are forecast to catch up to them and possibly pass them in revenue production in 2024. This is partly due to the rising price of oil, where both Coriolis and ultrasonic meters are heavily used. Even though water is diminishing in supply while demand for it is growing, the supply and demand picture for crude oil and refined products is at a totally different level. There is only a finite amount of crude oil and natural gas left in the ground. Even though there are new discoveries, the total amount of oil and natural gas is slowly being depleted, making it more valuable. By contrast, it is hard to imagine a situation where we run out of water as long as the oceans are around, even though portions of the earth experience droughts and scarcity of water.

Suppliers have made many advances in magnetic flowmeter technology to make them more accurate, reliable, and versatile. One advance is the introduction of two-wire magmeters. Four-wire magmeters have a dedicated power supply, while two-wire magmeters derive their power from the loop power supply. This can reduce wiring and installation costs.

A number of companies have introduced battery operated magmeters. This adds a new versatility to magmeters, since they can now be used in hard-to-reach places or places not powered by electricity. This is important since magmeters excel at measuring water flow, and water flow occurs in many remote places. This includes culverts, underground streams, streams in remote locations, and any other place that water flows and needs to be measured.

It should be noted that in some cases magnetic flowmeters measure flow through partially filled pipes or culverts. In these cases, the flow is due to gravity rather than due to pressure, as it is in full pipes. This flow is considered open channel flow, but it is still flow using magnetic flowmeters. Often an area velocity method is used to calculate the flow, which requires a knowledge of the area of the pipe, the velocity of the flow, and a level measurement.

Magnetic flowmeters are widely used for sanitary applications. This especially includes the food and beverage, pharmaceutical and life sciences, and chemical industries. Magnetic flowmeters used for sanitary applications have special liners that make them well-suited for this purpose. Coriolis meters are also used in sanitary applications, and both types of meters have meters with very small inner diameters, including 1/16th inch or less.

A tour of eight breweries and microbreweries in the Milwaukee, Wisconsin, area revealed that the two types of meters used most in these plants are magnetic and Coriolis. Magnetic flowmeters are widely used at the beginning of the brewing process. This especially applies to the malting, milling, mashing lautering, and boiling phases of the brewing process. As the brewing process proceeds and more precise measurement is necessary, some companies switch to Coriolis meters to achieve higher accuracy in measurement. At the end of the process magnetic flowmeters are also used for batching and filling.

Insertion magmeters are another innovation that enables these meters to measure flow in very large pipes at reduced cost. Because insertion magmeters measure flow at a point in the flowstream, they are not as accurate as full-bore meters. However, when high accuracy is not required, or cost is a factor, insertion magmeters can be a very viable choice.

Other major advances in magmeters were covered in Chapter 5. These include the move from AC to DC powered meters, and the continued need for AC powered meters for some applications. This chapter also covered the importance of liners, and the way that liners have made magmeters more adaptable to a wider variety of applications. Another important advance covered in Chapter 5 is the advance in signal processing and software that enables magnetic flowmeters to measure the flow of lower and lower conductivity fluids.

In summary, there have been many advances in a wide variety of areas for magnetic flowmeters. These advances are continuing along with the increased demand for accurate water measurement. Expect companies like Endress +Hauser, KROHNE, Emerson Rosemount, and Yokogawa to continue to bring out new magnetic flowmeters, and to continue to add new features, such as new communication protocols, to their magnetic flowmeters. The future for magnetic flowmeters looks bright and the demand for water measurement will continue to increase with growing populations and scarce resources in important areas.

THERMAL

Chapter 8 describes the development of thermal dispersion meters out of anemometer research. It also traces the development of thermal flowmeter companies that mostly started in the Silicon Valley area, then mushroomed into a worldwide business with suppliers elsewhere in the United States, in Europe, and in Japan. Thermal flowmeters are unique in that about 95 percent of them are used to measure gas flow, and the rest are used to measure liquid flows. This stems from the technology – heat inserted into the flowstream travels more quickly to a downstream sensor through gas than through liquid.

Accuracy is one area of concern for thermal flowmeters. When I initially proposed the distinction between new-technology and traditional technology (now conventional) flowmeters in a series of articles in *Control* magazine in 2001, I characterized thermal flowmeters as traditional meters because I didn't think that a flowmeter that couldn't measure at 1 percent accuracy deserved to be called a new-technology flowmeter. This has to do with the criterion "performs at a higher level than traditional (conventional) flowmeters."

Dr. John Olin, founder of Sierra Instruments, contacted me and challenged me on this classification. We had discussions that lasted over a period of six months. He sent me a paper he had written on the topic of thermal flowmeter performance. In the end, I was convinced that some thermal flowmeters under some conditions could measure with an accuracy of 1 percent. I relented, and wrote an article in *Control* magazine in October 2003 announcing the change in classification. Today thermal flowmeter suppliers are listing accuracy specifications at ±1 percent of reading and ±0.5 percent full scale. Accuracy is an area where thermal flowmeter suppliers have improved over time.

Another important area where thermal flowmeters have advanced has been in environmental applications. When Continuous Emissions Monitoring (CEM) came into effect in the early 1990s due to environmental legislation, thermal flowmeters found a new market for flow measurement. This gave a boost to the thermal flowmeter market, along with the averaging pitot tube and the ultrasonic markets. In 2008, when measuring greenhouse gas emissions became important as the Obama administration took power in the United States, thermal flowmeters once again experienced a boom in sales.

While environmental measuring was de-emphasized during the Trump presidency, addressing climate change is a major goal of the Biden administration. This includes further development of alternative fuels and energy sources such as biogas, biogas, fuel cells, and other environmentally friendly energy sources. Expect this trend to continue, since the move toward renewable energy will continue over the long term.

Flare gas and stack gas is another area where both insertion and inline thermal flowmeters are used. While some companies are being encouraged to use their gas rather than burn it off, there are many cases where safety requires the flaring of gas. Even though flaring may be reduced in some cases where the gas can be captured and used instead, there will still be a long-term need to flare gas and to measure stack gases. In this application, thermal competes with ultrasonic flowmeters and averaging pitot tubes using differential pressure measurement.

REDUNDANCY

Redundancy is becoming increasingly important in flow measurement. Redundancy has several benefits. In some cases it provides more accurate measurements. In other cases it provides a backup measurement when one flowmeter malfunctions or needs to be removed from service. Redundancy can also make it possible to check on the measurement accuracy of another meter.

In the Middle East and in other situations where oil and gas are measured for custody transfer purposes, sometimes two meters are run in series. They may be the same kind of meter, or they may be two different kinds of meters. For example, two ultrasonic meters might be run in series before a custody transfer measurement is made as a check on the accuracy of the total measurement. Alternatively, a turbine meter might be run in series with an ultrasonic meter to ensure measurement accuracy. This is sometimes done when the value of the product being measured is high and even small percentage mistakes have to be avoided.

Check metering is another form of redundancy. Clamp-on ultrasonic meters are sometimes used as check meters. Because they clamp onto the pipe, and do not require cutting the pipe for installation, clamp-on meters can easily be moved from one location to another. Even though clamp-on meters are not as accurate as some inline ultrasonic meters, this is a convenient and efficient way to check on the correct operation of another meter.

Emerson has introduced redundancy in vortex flowmeters. The redundant flowmeter has redundant, independent sensors and electronics. Transmitters are available with independent configurations. These flowmeters are used for Safety Instrumented Systems (SIS) and other applications where redundancy is critical. Emerson takes redundancy one step further with its Quad Vortex Flowmeter. Besides containing redundant sensors, this meter includes a fourth transmitter for more complete redundancy.

DUAL TUBE METER

The author of this book has received two patents that introduce redundancy into flow measurement. The idea behind the dual tube meter is to make the two flow measurements in smaller tubes inside the meter body. The results of these measurements are then sent independently to a transmitter, which blends them into a single measurement. Once the meter is calibrated and a K factor for the meter is determined, it is possible to get the highest accuracy from the two independent measurements. This also ensures that if one of the two sensors stops operating

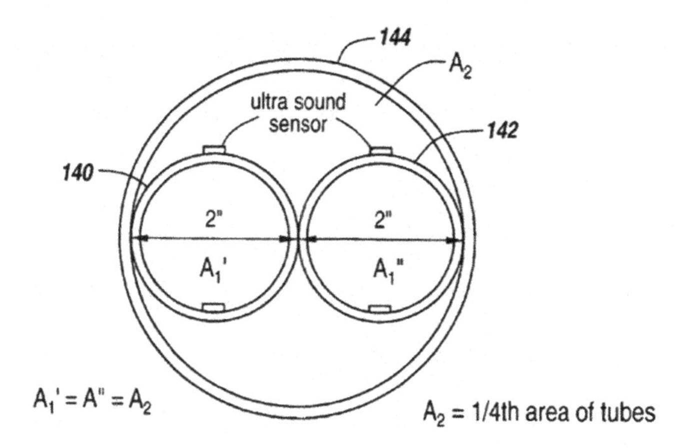

FIGURE 9.1 Patented dual tube flowmeter.

properly, the other sensor can continue to provide a flow measurement while the other sensor is replaced or repaired.

Different technologies can be used in either tube, or the same technology in both. Figure 9.1 shows a diagram of the patented ultrasonic dual tube flowmeter.

10 The Geometry of Flow

OVERVIEW

So far in this book we have looked at five types of new-technology flowmeters. Chapter 1 began with an overview of the entire book, and an explanation that this book is Volume I of a two-part series. Volume II covers conventional flowmeters and will do so in a way similar to the way new-technology flowmeters are covered in Volume I. Chapter 2 gave a definition of "flow," and explained some of the main concepts that apply to all types of flowmeters. Chapter 3 outlined a method for selecting flowmeters, using the theory of the paradigm case. The remaining chapters focused on individual types of flowmeters.

Chapter 4 explored some of the early patents on Coriolis flowmeters and talked about the role Jim Smith played in founding Micro Motion. It explains the prevailing theory about how Coriolis meters work, but it also questioned whether there is any real connection between how Coriolis flowmeters work and the theories and observations of Gustave Coriolis. Chapter 5 explained how magnetic flowmeters work and focused on two major magnetic flowmeter companies: Endress+Hauser and KROHNE. It also gave the growth factors for magnetic flowmeters. Chapter 6 discussed inline, clamp-on, and insertion ultrasonic meters. It focused on two major suppliers: SICK and Baker Hughes. This chapter gave growth factors for all three types of ultrasonic meters, including the use of inline meters for custody transfer of natural gas.

The subjects of Chapters 7 and 8 were vortex and thermal flowmeters. Vortex meters insert a bluff body into the flowstream and detect the frequency of the vortices this creates to determine flowrate. Chapter 7 describes an important part of this market: multivariable vortex meters. It identifies Jim Storer as the inventor of multivariable vortex meters and describes how he used Sierra Instruments as a distribution channel. This chapter also identifies growth factors for vortex meters. It describes VorTek and Yokogawa as significant suppliers of vortex meters. Chapter 8 describes how thermal flowmeters originated from anemometers, and how this technology developed in southern California in the 1970s and early 1980s. It focuses on Fluid Components International and Kurz Instruments as especially significant suppliers of thermal meters. It also describes some of the other companies that subsequently entered this market, including European and Asian companies.

WHAT IS FLOW

In Chapter 2, flow is defined as follows: "Flow is the continuous and uninterrupted motion of a fluid or a pattern of objects moving uniformly along a path in a direction."

DOI: 10.1201/9781003130017-10

The above definition captures the most common examples of flow, including river flow, the flow of a stream, flow of liquids within a pipe, and open channel flow. An open channel is the opposite of a closed pipe. In open channel flow, the flow moves by gravity, while in closed pipe flow, it is moved under pressure. Flow in partially filled pipes is considered to be open channel flow, since the liquid is not moving under pressure.

Since flow seems closely connected to continuity, it is worth looking at the idea of continuity. The number line itself is continuous, and yet many mathematicians view the number line as being made up of discrete points. What is confusing about this analysis is that points have no area. This idea follows the assumption that it is always possible to fit another point between any two points. However, if points have no area, meaning they do not have width but are essentially dimensionless, then 1,000 points or one million points will also not have area. Mathematicians compensate for this by using the idea of infinity, arguing that even if points have no area, surely infinitely many of them will have area. But infinity multiplied by 0 is still 0 and adding infinity to dimensionless points does not yield width, length, or the number line, which is continuous.

Part of the problem with this reasoning is that a line is made up of points. But if points lie on the line instead of being a part of the line, then the line can have continuity independent of the points. In fact, one way of conceiving of a line is as the path of a point in motion. Likewise, a plane is formed by placing a line in motion.

A LINE IS THE PATH OF A MOVING POINT

What is the relationship between points and a line? A line is the path of a moving point, as Aristotle says in De Anima 1:4. Likewise, a plane is the path of a moving line. A point and a line then are intimately related, but not in the way Euclidean geometry describes them as being related. A line is somewhat like the trail of a meteor, except that when we use the point of a pencil to draw a line, the line is static and remains visible.

POINTS LIE *ON* THE LINE, NOT *IN* THE LINE

Anyone who is aware of our language will realize that we speak of points being on a line more naturally than of points being in a line. The idea that points are in a line is more a result of mathematical analysis than of an understanding of mathematical language. But what is the difference between points being on a line and points being in a line?

Someone who is sitting on a fence is not part of the fence; instead, his or her body is physically touching the fence. But no one would think that a person sitting on a fence is part of the fence. Instead, the fence is made up of steel, wood, rocks, or some other material, depending on what type of fence it is. Likewise, a book lying on a table is not part of the table, although the book touches the table.

Intuitively, it makes sense to say that points lie on a line. When we draw a point on a line, typically the line is there first, and we physically mark the point on top of the line. We might us an "X" to mark the point ("X marks the spot"), with a round dot, or with a little perpendicular line. However it is marked, it would be unusual to conceive of this "X," round dot or perpendicular line as somehow being part of the line, while it is perfectly natural to think of the X, round dot, or perpendicular line as being on the line.

If points lie on the line rather than in the line, there is no need to introduce the concept of infinity to describe how many points there are. This is because a group of points lying on a line are not continuous; instead, they are instead a group of discrete points related by all being on the same line. It is the line that is continuous, but since discrete points are not part of a continuous line, they are not part of the continuous phenomenon.

HOW MANY POINTS LIE ON A LINE

If we cannot consider a continuous line as an infinite set of arealess points, how should it be analyzed? We have already said that points lie on the line but are not part of the line. In this analysis, points are discrete units that sit on the line. If the number line is being considered rather than a line that consists of the distance between two points, then these points can be considered as dimensionless in the Euclidean fashion. Euclidean mathematicians will consider these as representing an infinite number of points. In a non-Euclidean analysis, these points have area. This area will vary with the unit of measurement that is specified. Pursuing this non-Euclidean conception of a point as having area or width, no matter how small, it is always possible to redefine the width or area to any level of precision desired. For example, a point can be defined as one millionth of an inch, or one trillionth of an inch. Then there are a million points between 0 and 1, or a trillion points. These points are conceived as touching each other at the edges so there is no room for additional points once their size is determined. If a smaller or larger point is required, then the unit of measurement needs to be defined accordingly.

WHEN BOUNDARIES MATTER: DEFINING POINTS AND LINES

Sometimes, determining area by treating the boundaries of a figure as being a line with no width works well. There is often no issue about exactly where the boundary is. The boundary area between the end of one inch and the beginning of the next inch on a ruler is thought of as having no width, so that the question of where to begin and stop measuring does not arise. The lines on the ruler are lined up with the object being measured, and an inch is marked off.

Treating a boundary line as one with no width can also work in cases when the boundary line is so thin relative to what it borders that no purpose is served by treating the boundary line as having width. For example, a piece of rope that separates two tracts of land may be so thin relative to the size of the land that there is no point in specifying the boundary more precisely. Even if there is a small portion of land that lies directly on the boundary, this portion is so small that it can be ignored for the purposes of dividing the two tracts of land.

However, the boundary width may be significant when what lies on the boundary is important, or when the boundary line is large relative to the size of the marked area. For example, if gold or buried treasure lies on the boundary line between two properties, it may become important to specify which portion belongs to which property. With the center line on a highway, the line divides the two sides of highway and doesn't belong to either side. The line is significant in size relative to the width of the road, even though it may be only several inches wide, and is an example of a boundary line with important width. Cars are not allowed to cross this line in most circumstances except to pass another car on a broken line. A doorway between two rooms provides a similar example of a boundary with width. The area within the doorway typically doesn't belong to either room; it is there as a three-dimensional dividing line between the two rooms.

Football provides another example where physical boundaries have width. An American football field is marked off with nine 10-yard markers. Each yard line is several inches wide. If the football rests anywhere on these lines, it is "on the 10 yard line," for example. However, at the goal line, Euclidean geometry takes over again. To score a touchdown, the player with the ball must position the ball so that it breaks the plane of the goal line before he is "down." Here the inside edge of the goal line is treated as marking a vertical plane with no thickness that the ball must break for a goal to be scored.

In baseball, the situation is similar. Chalk lines are laid from home plate down the first and third base lines to distinguish fair from foul territory. These chalk lines are several inches wide. However, if a ball lands on the chalk line, it counts as a fair ball. It is only foul if it lands outside the chalk line. So, in this case, the chalk line is treated as an extension of fair territory. What is called the "foul pole" is really a "fair pole" since balls that hit it are considered to still be in play.

TWO CONCEPTIONS OF POINTS AND LINES

So, does it make sense to treat lines as having width? To answer this question, let us look at the function of measurement. Measuring the area or volume of an object is typically done to determine how many units of area or volume it contains. When someone is baking a cake, that person wants to know how many cups of flour he or she is putting in the cake. Likewise, quantities are important in commerce. A customer who buys a gallon of milk wants to know that she is getting one gallon, not some percentage of a gallon such as 3½ quarts. Two functions of measurement, then, are to specify quantities for practical matters such as recipes and to ensure that people get the advertised quantities of products.

If we treat lines as having no width, this may have no practical impact in some situations. To divide a piece of cake into two equal slices, it works to simply mark a line in the middle and physically divide the cake by cutting along the line. This act of division forces all particles of cake into one side or the other and creates two pieces of cake where formerly there was one. Of course, some crumbs may result that are particles of cake that didn't stick to one piece or the other, but these are insignificant by-products of cutting the cake in two pieces.

When the quantities are not being physically divided but only divided by a line, as in the border between two towns, the width of the line may be significant. In some cases, where the border is disputed, a no-man's-land may be specified to mark an area between two provinces or countries that belongs to neither one. For example, the Korean Demilitarized Zone (DMZ) is a 160 mile long and 2.5 mile wide border between North and South Korea. It was created in 1953 as part of the armistice agreement between South and North Korea. It is a buffer zone between the two countries, it is roughly located at the 38th parallel, and is not part of either country.

In mathematical examples, theoretical problems arise in specifying the exact border or boundary of geometrical objects. (The terms "border" and "boundary" are synonyms, except that "boundary" is often used to refer to a dividing line between two areas, including countries or tracts of land. The term "border" is often used to refer to the edge of a geometric or physical object when it is not bounded by another similar area.) It is reasonable to wonder, for example, whether the area of a circle only includes the area within the circle or whether it also includes the border of the circle. This is especially true since the area of a circle is, by conventional mathematics, specified by an irrational number, so that clarifying exactly what "area of a circle" is might shed some light on our inability to specify this area with rational numbers. The same question could be asked about rectilinear figures, although the corresponding question involving irrational areas does not arise for them.

WHAT IS A LINE

Whether a line has width depends on what we mean by "line." Adolf Grunbaum comments on this issue. Speaking in the context of a discussion of Cantor's set theory, he says:

> No clear meaning can be assigned to the "division" of a line unless we specify whether we understand by "line" an entity like a sensed "continuous" chalk mark on the blackboard or the very differently continuous line of Cantor's theory. The "continuity" of the sensed linear expanse consists essentially in its failure to exhibit visually noticeable gaps as the eye scans it from one of its extremities to the other. There are no distinct elements in the sensed "continuum" of which the seen line can be said to be a "structured aggregate." From "A Consistent Conception of the Extended Linear Continuum as an Aggregate of Unextended Elements" *Philosophy of Science 19*(4) (October 1952): pp. 288–306.

Does the idea that a line has no width make sense? This is Euclid's definition, who defined a line as a "breadthless length." (See Definition 2 in Euclid's Book One.) This idea is also consistent with Plato's view that mathematics is about ideal, abstract objects, not about physical lines and curves. Someone who draws a rectangle and calculates the area as length × width ($l \times w$) will not retract his statement if the lines are not completely straight or if the length and width do not form an exact 90° angle. The equation is about an abstract set of lines and relations that the drawn figure represents, not about the physically drawn rectangle.

The idea that lines do not have width was Euclid's view, it was Plato's view, and it is the established view of Euclidean geometry as it is taught today. Instead of arguing against this view, which most people take for granted, I am presenting an alternative conception that proceeds from a different set of assumptions. These are simply two different ways of analyzing the fundamental concepts of point, line, and area, rather than being competing geometries. I propose a geometry in which points have area and lines have width. I believe that this geometry more closely captures how we actually conceive of points and lines in certain circumstances, as discussed above.

The idea that a line has width follows from Aristotle's definition of a line as the path of a point in motion. The width of a line equals the diameter of the point used to draw the line. This does not mean we have abandoned Plato's view that geometry is about idealized objects rather than physical drawings. It only means that the abstract lines represented by physically drawn lines are conceived as having width. I propose to call this Wide Line Geometry.

WIDE LINE GEOMETRY

Aristotle's definition of a line as "the path of a moving point" seems preferable to Euclid's definition of a line as a "breadthless length." Of course, this leaves the concept of a point undefined.

It is possible to treat a line as having no width for measurement purposes. Any drawn line, no matter how thin, has some width. The width of a drawn line is similar to the duration of a unit of time. It is not possible to specify any time period that does not have duration. One hour, one minute, one second, one millisecond, and one nanosecond all have some duration. Of course, sometimes it is convenient for measurement purposes to consider a unit of time such as a second as being "a point in time" with no duration. Likewise, it is useful to treat both points and lines as being dimensionless for some measurement purposes.

While drawn lines have width, Wide Line Geometry is not about drawn lines. Wide Line Geometry retains Plato's view that drawn mathematical points and lines are about ideal objects that are represented by these drawn points and lines. The merit of Wide Line Geometry is that it coincides more closely with the way we actually treat lines in certain situations. The yard lines in an (American) football field, the chalk line on the edge of a baseball field, and the lines dividing the highway into two sides are three common examples in which we treat lines as having width. In fact, Wide Line Geometry comes much closer to capturing the way we actually treat points and lines than does Euclidean geometry with its "breadless lengths."

LINES, AND THE NATURAL AND REAL NUMBER LINES

Some concepts are so fundamental that it is difficult to give a meaningful definition of them in more intuitive terms. Euclid's first definition in Book One of Euclid's *Elements* is "A point is that which has no part." This definition could be criticized on the grounds that objects cast from a mold do not have parts but are not considered points. Spoons, candlesticks, statues, pots, bullets, and other objects cast from molds do not have parts. While Euclid is trying to define a mathematical concept, not one that applies to physical objects, his definition still applies to these physical objects.

Euclid's definition of a line as a "breadthless length" uses the concept of length to define what a line is, but the concepts of line and length seem very closely related. Every line has length, and Euclid makes it a matter of definition. As far as the "breadthless" part goes, this definition cannot apply to physical or drawn lines, since every drawn line has breadth, or width. Euclid is referring to a line as an abstract mathematical concept that a physical or drawn line represents – an abstract line with no width.

INFINITY AND THE NUMBER LINE

What about infinity? Aren't there infinitely many numbers? I prefer to say, along with Descartes, that there are indefinitely many numbers. This means there is an unlimited number of numbers, but they do not exist as a completed set, as set theory proposes. This means that we will never run out of numbers, but there are not infinitely many of them. As for irrational numbers such as the square root of 2 and π, the need for these only arises out of implicitly contradictory assumptions. The need for π only arises because we are attempting to determine the exact number of squares that fit into a circle. It is equally logical to say that there is no such number instead of calling it π and attempting to define it as a series of nonrepeating decimals that goes to infinity.

MAKING A MEASUREMENT REQUIRES A UNIT OF MEASUREMENT AND A DEGREE OF PRECISION

When a measurement is made, a unit of measurement is either explicitly specified or understood. If I say this stick measures four, I have not given a measurement until I specify what unit of measurement I am using. I might mean four centimeters, four inches, four feet, four yards, or four meters. Just giving a number doesn't state a measurement apart from a unit of measurement. Often the context makes this clear, though often the unit of measurement needs to be explicitly stated.

A degree of precision is a second requirement for a measurement. This may seem less obvious, since degrees of precision are not always specified, but a few examples should make this clear. When the weatherman gives the weather, it is almost always in whole degrees. If he predicts temperatures of 65–70°F, people are not going to expect him to predict to the tenth of a degree, like 65.5–70.3°F. It is clear just by using whole numbers that the forecasts are stated in terms of whole degrees and not in terms of tenths of a degree.

If someone gives the distance from the earth to the sun as being 93 million miles, no one is likely to demand that this number be translated into inches or feet. Furthermore, it is generally understood when this number is given that it is being given in round whole numbers, and that the number of 93,000,000 miles is most likely rounded up or down from a more precise number. Given the size of the earth, the fact that it is constantly in motion and is sometimes closer to the sun than at other times, the added "precision" of feet or inches doesn't add any accuracy to this measurement.

Precision in time is also important in a parallel way. If I ask what time it is and you say, "It's noon," I am not likely to object if it is only one minute before noon. However, sometimes we need to know exactly to the minute or even to the second what the time is. This precision is often critical in making appointments and in sporting events. Even so, the time is not usually stated beyond the precision of minutes except in scientific measurements and in certain sporting events. An example is the last minutes of many professional basketball games when time is measured in tenths of a second.

There are 1,440 minutes in a day and 86,400 seconds. This doesn't vary, except in other systems such as decimal time, but just as points have area, seconds have duration. It is easy to think that a second is a "point" in time, but a second lasts a second, a minute a minute, and an hour an hour. There is no such thing as a unit of time without duration. Otherwise, there would be infinitely many points in time, which there are not.

Sometimes an hour or even a minute can seem like an eternity. I experienced this while waiting for my PhD dissertation committee to come back with their verdict after my oral exam (fortunately, it was positive, or I probably wouldn't have written this book). While we use the expression "point in time," units of time are very much like points on a line. Just as units of time have duration, so points on a line have area. And the amount of area depends on the unit of measurement.

LENGTH IN FLOW MEASUREMENT: DOES A PIPE CIRCUMFERENCE HAVE WIDTH

The concept of length is critical in flow measurement. Fluids typically travel in round pipes, and flowrate is measured by the classic equation, which is explained in Chapter 2:

$$Q = V \times A$$

Here Q is flowrate, V is velocity, and A is cross-sectional area. The area of a pipe is typically determined by the equation πr^2. Here r is the radius, which is one-half of the diameter of the pipe. The inner diameter of a pipe is a straight line that runs from one side of the pipe to the other through the center of the pipe. Radius is a measurement of length and is fundamental to measuring flowrate.

It is important to distinguish between the inside diameter (ID) of a pipe and its outside diameter (OD). When measuring flow, it is the inside diameter of a pipe that is relevant to determining flowrate. While it is common in Euclidean geometry to think of the circumference of a circle as having no width, this is not how pipes are in reality. While pipes are round, for the most part, their circumference has width. The circumference of a pipe is its wall thickness. Wall thickness is important for some types of flow measurement, such as clamp-on ultrasonic flowmeters.

One common problem in determining an accurate flowrate is that buildup can occur on the inside of pipes. This impacts the accuracy of flowrate measurement since it decreases the inside pipe diameter. For example, ultrasonic flowmeter accuracy can be affected if pipe buildup occurs, since it reduces the length that the ultrasonic signal travels. Clamp-on ultrasonic meters that send an ultrasonic signal through the pipe wall can also be negatively impacted by pipe buildup. The pipe wall may already cause the signal to be attenuated, and pipe buildup can further attenuate the signal.

CIRCULAR GEOMETRY

If you studied mathematics or geometry in high school or college, you probably learned the following formula for the area of circle:

$$A = \pi \times r^2$$

Here A is the area of the circle, while r is the radius of the circle. The number π, which represents the ratio of the circumference to the diameter of the circle, is an irrational number that has never been completely specified.

What is this formula actually asking us to do? If we look at the geometry of this formula, it looks like the diagram in Figure 10.1.

The value r^2 gives the geometric area of the square in the above diagram. The formula for the area of a circle, then tells us that π squares with sides equal to radius r fit into the area of a circle with radius r.

What, if anything, is the problem with this formula? And why do we need to have π in the formula? The reason for π is that there is no definite number of times that a square can fit inside a circle. It is often said "You can't fit a square peg into a round hole." This common saying reflects the insight that the area of a square cannot be used as a unit of measurement for circular area. Since there is no definite number of times that a square will fit inside a circle, the value π has to be included to create a usable formula for circular area.

The relation between circular area and square area is that they are incommensurable. What this means is that they cannot be both measured exactly using the same standard or unit of measure. Straight lines and squares work fine for squares and rectangles, but they do not allow us to provide exact values for the areas of circles.

If the areas of squares and circles cannot be measured exactly using the same unit of measurement, we have several choices. One is to continue as we are, using

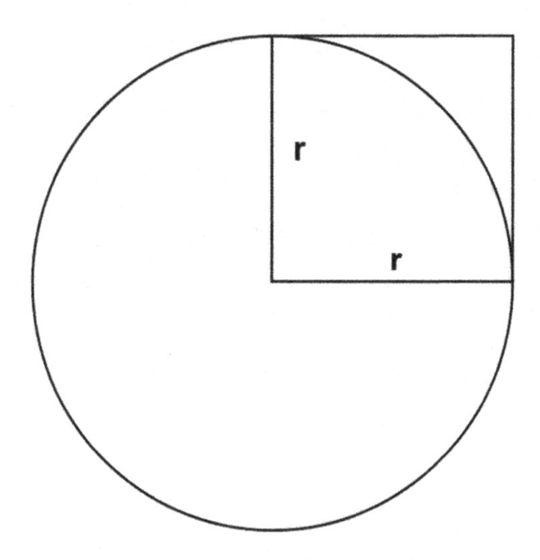

FIGURE 10.1 The value r^2 as a unit for measuring the area of a circle.

square area as the unit of measurement for circular area. This has the advantage of familiarity, provided we don't mind using π. A second alternative is to use a different unit of measurement for circular area. This is the alternative I would like to suggest here.

An Alternative Unit of Measure

As an alternative unit of measure for circular area, I suggest the round inch. A round inch is a circle with a diameter of one inch. If we use the round inch as the unit of measure for circular area, this unit of measure looks like the drawing in Figure 10.2.

In Figure 10.2, each of the two smaller circles is equal to $A/4$, where A is the area of the circle. Each small circle has a diameter of one inch, and so is equal to one round inch. The area of the large circle is equal to $(\text{diameter})^2$. This value is equal to 2^2, which is equal to 4. So, the circle above has an area of four round inches.

Circular Mils

A similar approach to this already exists for measuring the area of round wire. In order to avoid using decimals, the area of round wire is often measured in circular mils. The area of a circular mil is $(\text{diameter})^2$, so the formula is as follows:

$$A = (\text{diameter})^2$$

A mil is equal to 1/1,000th of an inch (0.001 inch). A circle that has as one mil as its diameter has an area of $1^2 = 1$. A circle that has a diameter of four mils has an area of $4^2 = 16$ circular mils.

What is the relation between a round inch and circular mils? A round inch has a diameter of one inch. Since a mil is 1/1,000th of an inch, a round inch has a

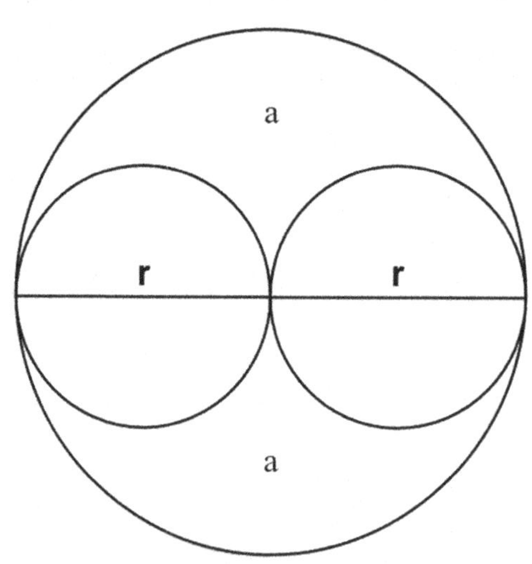

FIGURE 10.2 The round inch is the unit of measurement for circular area.

diameter of 1,000 mils. This means that a round inch has an area of $1{,}000^2$ circular mils, or 1,000,000 circular mils. So, any area that can be measured in round inches can be measured in circular mils and vice versa.

Application

What is the application for circular geometry? Circular geometry can be used anywhere someone wants to measure circular area. It is true that many buildings and other structures are either square or rectangular. This is an example of geometry influencing architecture. Because most of the geometry that is taught in schools is so uncompromisingly linear, meaning that it is based on straight lines, squares, and rectangles, many of the buildings and other structures that are created using this geometry reflect its underlying linear nature. On the other hand, if we look in nature, we find a wide assortment of waves, curves, circles, and other nonlinear geometric shapes. The world is round, even though it looks flat, and many natural shapes are nonlinear as well.

One area that circular geometry has an application is in flow measurement. Pipes are round, and it is often necessary to determine the area of a pipe in order to determine volumetric flow. The formula that is usually used is the following one:

$$Q = A \times v$$

In the above formula, Q is equal to volumetric flow, A is the cross-sectional area of the pipe, and v is the average velocity of the fluid. It is in providing the cross-sectional area of a pipe that the value of π is used in calculating flowrate. If this area is provided in round inches rather than square inches, flowrate can be calculated without the use of π.

Round inches can generally be substituted for square inches in geometry when calculating the area of circles. Just as circular mils are used to measure the area of wire, so round inches can be used to measure the areas of circles. Of course, just as there is no exact way to measure circular areas in terms of straight lines, there is no way to exactly measure the area of squares and rectangles using circular geometry. Just as a hammer is used for nails, and a screwdriver is used for screws, so each type of geometric structure requires its own geometry.

Sensing and Measuring

One important distinction that underlies the discussion in this book is the distinction between sensing and measuring. Flowmeters may not sense flow directly, but they all contain sensors that sense some parameter that is associated with flowrate. For example, magnetic flowmeters have electrodes that sense the voltage created when a conductive liquid passes through a magnetic field and use this information to compute flowrate. Coriolis flowmeters sense the impact of fluid momentum on a vibrating tube and use this information to determine mass flow.

Flowmeters contain sensors, but they are also measuring devices. A measuring device contains a unit of measurement, and this unit of measurement is used to determine a quantity that determines how much of that unit something has. For example, if a flowmeter has gallons per minute as a unit of measurement, it uses information from the sensor to determine how many gallons of a fluid are passing through a pipe in a given period of time. Just as thermometers have Kelvin, Celsius, and Fahrenheit scales, so flowmeters have different units in which they measure. Some units are Imperial (American) and others are metric (SI). The units can often be programmed into the flowmeter.

What Is a Sensor

In order for something to be a sensor, it must respond in a predictable way to some property or parameter in the object it is sensing. This information must be used to detect the presence of some quality in the object. This means there must be some interaction between the object being sensed and the sensor. However, not all forms of interaction count as sensing. A stick of wood placed into a river isn't sensing the flow; it just deflects it. A bar of steel does not sense light even if it reflects it because the reflected light is not used to determine the presence of a quality in the bar of steel; the light is simply reflected.

Sensors that are sensing an object share the following characteristics:

1. They are made from some kind of physical material.
2. This material is sensitive to changes in the qualities of a physical property of the object.
3. The sensing material or property responds to changes in the qualities of a physical property in the object in a predictable way.
4. A converter, transducer, or transmitter takes these changes and converts them to a reading of flow, temperature, pressure, or whatever the sensed variable is.

What kind of relationship has to exist between a sensor and the property or sensed object? It has to be a predictable relationship, at the very least. A mercury thermometer whose readings vary wildly and inconsistently with the temperature is not sensing it, even if it is responding to it. An outside thermometer that reads 70 degrees when it is 30 degrees outside and 86 degrees when it is 40 degrees is not sensing the temperature, but instead responding in a seemingly arbitrary way to it. The idea of sensing contains the idea of truth, so that a sensor portrays an objective value, within certain bounds of correctness. Even so, a thermometer that is a few degrees off is still sensing the temperature even if it is not completely accurate.

What is the relation between a sensor and the property or sensed object that makes it a sensor? Because of the element of truth or accuracy that is implied in saying that something is a sensor, they must be related in a predictable way. This means that the sensor is following a rule in the presence of the sensed object or property. This rule may not always be known, but it must exist. It is the rule that formulates the predictable relation between the sensor and the property or sensed object. In the case of a mercury thermometer, when the mercury is at a certain height, it reads 50 degrees, and when it is at a different height it reads 80 degrees.

The height of the mercury in response to temperature depends on the expansion powers of mercury.

This enables us to formulate a fifth principle relating to sensors:

> When a sensor exists and senses the presence of a property or an object, it is following a rule that formulates the relation between the senor and the sensed object or property. This rule may or may not have been explicitly formulated.

Sensors and the Mind

In Volume II of this series, *Conventional Flowmeters*, we will examine the three types of sensors, discuss the roles of transducers and transmitters, and further explore the relation between sensing and measuring.

After this, we will discuss what electronic and mechanical sensors tell us about biologic sensors. The mind is a biological sensor, so we will explore what we can learn about our five senses and the mind from studying electronic and biological sensors.

Bibliography

Chapter 1

Mark Patinkin, Investigating the Conspiracy: What Happened to 1-Pound Can? (Deseret News, May 1, 1989)

Chapter 2

Endress+Hauser, Flow Handbook (Reinach: Endress+Hauser AG, 2006)

IDC Technologies, Fiscal Flow Metering Equipment (West Perth, Australia: IDC Technologies Pty Ltd., 2011

Industrial Flow Measurment, David W. Spitzer (Research Triangle Park, North Carolina: Instrument Society of America, 1990)

E. Loy Upp, Fluid Flow Measurement (Houston, Texas, Daniel Industries, Inc., 1993)

Chapter 3

Flow Research, Inc., Volume X: The World Market for Flowmeters, 8th Edition (Wakefield, Massachusetts, Flow Research, Inc., January 2022)

Chapter 4

Flow Research, Inc., The World Market for Coriolis Flowmeters, 6th Edition (Wakefield, Massachusetts, Flow Research, Inc., September 2020)

Jesse Yoder, The Coriolis Effect (Nashville, Tennessee, Flow Control Magazine, Grand View Media Group, November 2011)

Chapter 5

Flow Research, Inc., The World Market for Magnetic Flowmeters, 7th Edition (Wakefield, Massachusetts, Flow Research, Inc., May 2022)

Chapter 6

Flow Research, Inc., The World Market for Ultrasonic Flowmeters, 6th Edition (Wakefield, Massachusetts, Flow Research, Inc., May 2021)

Chapter 7

Flow Research, Inc., The World Market for Vortex Flowmeters, 6th Edition (Wakefield, Massachusetts, Flow Research, Inc., January 2019)

Chapter 8

Flow Research, Inc., The World Market for Thermal Flowmeters, 2nd Edition (Wakefield, Massachusetts, Flow Research, Inc., January 2018)

Chapter 9

Jesse Yoder and Richard E. Morley, The Tao of Measurement (Research Triangle Park, North Carolina, International Society of Automation, 2015)
Euclid, The Thirteen Books of The Elements (Toronto, Canada, General Publishing Company, Ltd. 1956)
Al Shenk, Calculus and Analytic Geometry (Glenview, Illinois, Scott Foresman and Company, 1984)

Chapter 10

Jesse Yoder and Richard E. Morley, The Tao of Measurement (Research Triangle Park, North Carolina, International Society of Automation, 2015)

Index

Printed in the United States
by Baker & Taylor Publisher Services